7th International Conference on Clean and Green Energy (ICCGE 2018)

IOP Conference Series: Earth and Environmental Science
Volume 154

Paris, France
7 – 9 February 2018

ISBN: 978-1-5108-6337-8
ISSN: 1755-1307

Printed from e-media with permission by:

Curran Associates, Inc.
57 Morehouse Lane
Red Hook, NY 12571

CURRAN ASSOCIATES INC.
proceedings
.com

Some format issues inherent in the e-media version may also appear in this print version.

Printed by Curran Associates, Inc. (2018)

Institute of Physics
Dirac House, Temple Back
Bristol BS1 6BE UK

Phone: 44 1 17 929 7481
Fax: 44 1 17 920 0979

techtracking@iop.org

Additional copies of this publication are available from:

Curran Associates, Inc.
57 Morehouse Lane
Red Hook, NY 12571 USA
Phone: 845-758-0400
Fax: 845-758-2633
Email: curran@proceedings.com
Web: www.proceedings.com

Table of contents

Volume 154

2018 7th International Conference on Clean and Green Energy-ICCGE 2018

7–9 February 2018, Paris, France

Accepted papers received: 4 May 2018
Published online: 22 May 2018

Preface

2018 7th International Conference on Clean and Green Energy-ICCGE 2018

Conference Committee

Peer review statement

Papers

Renewable Energy and Energy Management

GIS-based approach for the evaluation of offshore wind power potential for Gujarat

Dhrumin Patel, Garlapati Nagababu, Nishil Radadia, Sohil Parsana, Mohak Sheth and Nisarg Sheth.....1

Climate conditions of the "El Niño" phenomenon for a hydro-eolic complementarity project in peru

Leonardo Castillo N, Arturo Ortega M and Jaime E Luyo.....9

Promotion of renewable energy in some MENA region countries

K. Abdeladim, S. Bouchakour, A. Hadj Arab, S. Ould Amrouche and N. Yassaa.....17

Decentralized and cost-effective solar water purification system for remote communities

Hafiz M. Abd-ur-Rehman, Sehar Shakir, Atta-ur-Razaq, Hamza Saqib and Saad Tahir.....23

Performance degradation of photovoltaic modules at different sites

A Hadj Arab, I Hadj Mahammed, S Ould Amrouche, B. Taghezouit and N. Yassaa.....29

The impact of predicted demand on energy production

I. El kafazi, R. Bannari and My. O. Aboutafail.....37

Building and Energy

Designing a dormitory with emphasis on renewable energy

F. Daneshvar Tarigh, A. Daneshvar Tarigh and F. Habib.....46

Comparative analysis of heat dissipation panels for a hybrid cooling system integrated in buildings

A Zuazua-Ros, JC Ramos, C Martín-Gómez, Tomás Gómez-Acebo and A Pisano.....59

Passive designs and renewable energy systems optimization of a net zero energy building in Embrun/France

F Harkouss, P H Biwole and F Fardoun.....67

A comparative study of fluorescent and LED lighting in industrial facilities

C Perdahci PhD, H C Akin BSc and O Cekic Msc.....74

Illumination of the marmaray project

C Perdahci, D Yuce and R. A. Aynaci.....83

Power System and Automation

Viability of using energy storage for frequency regulation on power grid

Y S Lim, L C Hau, K Y Loh, K Y Lim, P F Lyons and P C Taylor.....90

Analysis of turbine-grid interaction of grid-connected wind turbine using HHT

A Chen, W Wu, J Miao and D Xie.....97

Simulation of load-sharing in standalone distributed generation system

Titus O Ajewole, Robert P M Craven, Olakunle Kayode and Olufisayo S Babalola.....103

Using particle swarm optimization to enhance PI controller performances for active and reactive power control in wind energy conversion systems

M Taleb, M Cherkaoui and M Hbib.....111

Maximum power point tracking techniques for wind energy systems using three levels boost converter

Cuong Hung Tran, Frédéric Nollet, Najib Essounbouli and Abdelaziz Hamzaoui.....118

Environmental and Chemical Engineering

Environmental profile evaluations of piezoelectric polymers using life cycle assessment

M. A. Parvez Mahmud, Nazmul Huda, Shahjadi Hisan Farjana and Candace Lang.....127

Application study of Bio-FGD based on environmental safety during the coal combustion

Pin Zhang.....135

Environmental impact assessment of european non-ferro mining industries through life-cycle assessment

Shahjadi Hisan Farjana, Nazmul Huda and M. A. Parvez Mahmud.....144

Initial solubility & density evaluation of Non-Aqueous system of amino acid salts for CO_2 capture: potassium prolinate blended with ethanol and ethylene glycol

Ghulam Murshid and Sahil Garg.....151

Pyrolysis of polyethylene terephthalate containing real waste plastics using Ni loaded zeolite catalysts

M. Al-asadi and N. Miskolczi.....158

Co-pyrolysis of biomass and plastic wastes: investigation of apparent kinetic parameters and stability of pyrolysis oils

B. Fekhar, N. Miskolczi, T. Bhaskar, J. Kumar and V. Dhyani.....165

Transient analysis of a solid oxide fuel cell stack with crossflow configuration

P Yuan and S F Liu.....172

IOP Conf. Series: Earth and Environmental Science **154** (2018) 011001 doi:10.1088/1755-1315/154/1/011001

PREFACE

It is our great pleasure to welcome you to 2018 7th International Conference on Clean and Green Energy (ICCGE 2018) which will be held in NOVOTEL PARIS CRETEIL LE LAC, Paris, France during February 7-9, 2018. ICCGE 2018 is dedicated to issues related to Clean and Green Energy.

The major goal and feature of the conference is to bring academic scientists, engineers, industry researchers together to exchange and share their experiences and research results, and discuss the practical challenges encountered and the solutions adopted. Professors from France are invited to deliver keynote speeches regarding latest information in their respective expertise areas. It will be a golden opportunity for the students, researchers and engineers to interact with the experts and specialists to get their advice or consultation on technical matters, sales and marketing strategies.

This proceeding presents a selection from papers submitted to the conference from universities, research institutes and industries. All of the papers were subjected to peer-review by conference committee members and international reviewers. The papers selected depended on their quality and their relevancy to the conference. The volume tends to present to the readers the recent advances in the field of clean and green energy and various related areas, such as energy science and technology, environmental science and engineering, motivation, electrical engineering and automation, the development and utilization of resources, theory and practice of sustainable development, etc..

We would like to thank all the authors who have contributed to this volume and also to the organizing committee, reviewers, speakers, chairpersons, sponsors and all the conference participants for their support to ICCGE 2018.

Prof. Hartmut Hinz
University of Applied Sciences, Frankfurt, Germany
February. 23, 2018

IOP Conf. Series: Earth and Environmental Science **154** (2018) 011002 doi:10.1088/1755-1315/154/1/011002

Conference Committee

Conference Chair

Prof. Hartmut Hinz, University of Applied Sciences, Frankfurt, Germany

Prof. Michael Schmidt, Offenburg University of Applied Sciences, Germany

Program Chair

Prof. Marco Casini, SAPIENZA University of Rome, Italy

Assoc. Prof. Mejdi Jeguirim, University of Haute Alsace, France

Technical Committee

Prof. Ahmed A. Hossam-Eldin, Alexandria University, Egypt

Prof. Poul Alberg Østergaard, Department of Planning, Aalborg University, Denmark

Prof. Yusuf Ali Kara, Bursa Technical University, Turkey

Prof. Benatiallah Ali, Adtat University, Algeria

Prof. Hedayat Omidvar, National Iranian Gas Company, Iran

Prof. Saad A. EL-Syaed Hamad, Zagazig University, AlSharkia, Egypt

Prof. EL OUALIDI Jalal, University Mohammed V of Rabat (Institut Scientifique), Rabat, Morocco

Prof. Dilip Kumar De, Covenant University, Nigeria

Prof. BENATIALLAH Ali, Africane University of Adrar, Algeria

Prof. Ali CHEKNANE, Amar Telidji University of Laghouat, ALGERIA

Prof. Yun Seng Lim, Universiti Tunku Abdul Rahman, Malaysia

Prof. Mohammad_K. Ghassem-Alaskari, Petroleum University of Technology and Science, Iran

Prof. María José Cocero, Valladolid University, Spain

Prof. Xin Li, Zhejiang University, China

Assoc. Prof. Juan Carlos Ramos, University of Navarra, España

Assoc. Prof. Chao-Tsung Ma, National United University, Taiwan

Assoc. Prof. Eid M. Al-Mutairi, King Fahd University of Petroleum & Minerals, Saudi Arabia

Assoc. Prof. Omid TAVAKOLI, University of Tehran, Iran

Assoc. Prof. Hasan AYDOGAN, Selcuk University, Konya Turkey

Dr. Nada Alrikabi, University of Baghdad

Dr. Sreekanth K. J., Kuwait Institute for Scientific Research (KISR), Kuwait

Dr. Yang Du, Illinois State Geological Survey, USA

Dr. SHIBANI CHAUDHURY, Department of Environmental Studies, India

Dr. Nurcan Kilinc Ata, Osmaniye Korkut Ata University, Turkey

Dr. Behzad Alipoor, Islamic Azad University, Tehran, Iran

Dr. Rizwan Muhammad, University of Haripur, Pakistan

Dr. Suaad Jaber Laffta, Baghdad University, Iraq

Dr. Naila Rozi, Sindh Madressatul Islam University, Karachi, Pakistan

Dr. Ashwin Gaikwad, Visvesvaraya National Institute of Technology, India

Dr. Cesar Martín-Gómez, Universidad de Navarra, Spain

Dr. Norbert Miskolczi, University of Pannonia, Hungary

Peer review statement

All papers published in this volume of *IOP Conference Series: Earth and Environmental Science* have been peer reviewed through processes administered by the proceedings Editors. Reviews were conducted by expert referees to the professional and scientific standards expected of a proceedings journal published by IOP Publishing.

IOP Conf. Series: Earth and Environmental Science **154** (2018) 012001 doi:10.1088/1755-1315/154/1/012001

GIS-based approach for the evaluation of offshore wind power potential for Gujarat

Dhrumin Patel[1], Garlapati Nagababu[2], Nishil Radadia[1], Sohil Parsana[1], Mohak Sheth[1] and Nisarg Sheth[1]

[1] Student, Pandit Deendayal Petroleum University, Gujarat, India
[2] Assistant Professor, Pandit Deendayal Petroleum University, Gujarati, India

Abstract. In the current global scenario, India is increasing its focus towards the methods to enrich the benefits of non-renewable energy sources as much as possible due to their key advantage of having low carbon footprint. India has already emerged as a key global player in on-shore wind energy and to achieve its annual wind energy production demand of 50 GWh, avenues other than current options have been researched on. Offshore wind energy has experienced remarkable growth worldwide but has not yet been harnessed sufficiently in India, despite addressing many of environmental and economic concerns. The present study focuses on offshore wind resource assessment on Indian exclusive economic zone (EEZ) around Gujarat region. The geographical information system (GIS) methodology has been used to develop maps of wind speed, power density and capacity factor maps. Further, careful consideration has been accorded for expulsion of marine protected areas, shipping transportation lines, fishing zones, and migratory bird movements. The resultant available area has been considered for annual energy production considering data from Siemens Wind Turbine 3.6. The results obtained shows that offshore wind energy can offset twice the annual energy demand of entire country with a potential energy production of more than 2580 TWh.

1 Introduction

The Indian wind industry is nearly thirty years old, and now holds the 4th position in the world with installations of over 31 GW. According to the Conference of Paris (CoP-21), India has committed to produce 40% of its electricity from non-renewable sources by 2030. India committed to commission 60 GW of power production capacity from wind resources by 2022 [1].

Onshore wind turbines are associated with some negative environmental implications such as visual and noise impacts, impacts on birdlife and other types of environment impacts [2]. This decelerates future development of onshore turbine installations and makes it difficult for energy planners to localize suitable sites. The wind resource at offshore site is generally larger than at geographically nearby onshore sites, which can offset their higher installation, operation and maintenance costs. Successful offshore wind energy development relies to some extent on accurate prediction of wind resources, but since installing and operating a meteorological in situ is expensive, prospective sites must be carefully evaluated.

Gujarat has a long continental shelf over 50-60 per cent of its total coastline wherein the slope of the sea is gradual thereby setting up of offshore wind power is conducive. In addition, southwest winds come all over Gujarat from the Arabian Sea, thereby providing unhindered wind for at least six months. Hence, it has advantage compared to other states and huge potential to produce power from offshore winds from its coast [3]. The average wind speed in Gujarat recorded by the National

Renewable Energy Laboratory (NREL) is 8m/s and is sufficient speed to generate electricity by a wind turbine [4].

Carrying out wind measurement for several years is considered as the most suited approach to estimate wind energy potential of a region [5]. But to cover a large offshore area, it would be infeasible to perform that sort of measurement by add-on meteorological or measurement stations. Hence, depending on the situation, different methodologies are applied to assess wind energy accordingly. Moller [6] presented a model called Spatially Continuous Resource Economic Assessment Model for Offshore Wind Energy (SCREAM-offshore wind) based on GIS. It is not suitable for actual setting, as it does not precisely direct to the exact location. Dhanju et al.[7] presented a preliminary analysis for off shore wind potential on east coast of US by mapping wind speed and bathymetric data an exploring the possibility of setting up an offshore wind farm or multiple farms.

Thus, in lieu of offshore meteorological towers, some offshore wind resource studies extend land-based measurements to the areas over the ocean via modeling techniques [8], [9]. Others have characterized the resource by using offshore wind measurements from buoys or satellites and extrapolating those measurements to wind turbine hub height [7]. Furthermore, data collection on competing uses, such as environmental habitats, shipping and geologic conditions of the ocean is less complete than on land. Reference from the bathymetry data, shipping lines, fishing zones and protected marine areas have played a key role to obtain the useful area for the installation of the offshore wind turbines by excluding the other non-useful areas of the sea.

In this paper, a comprehensive wind power assessment has been performed to assess power generation potential in offshore region of Gujarat state in India. The results provide an estimate of Gujarat's feasible offshore wind energy resource. Offshore area of Gujarat was being assimilated based on bathymetric data to determine potential wind turbine foundation area. Thereafter this data was reanalyzed for potential wind area for power generation using wind speed data. The novel concept of exclusion of potential areas considering the detrimental effects such as environmental factors, shipping lines, marine protected area and fishing zones which have been uniquely identified using various sources. These factors are being subtracted to reanalyze the actual wind power potential. However, the results of wind energy harnessing potential obtained from this analysis are still greatly optimistic. We hope that this study will encourage further implementation of the offshore wind power resource and its ability to provide renewable electricity for coastal regions.

2 Methodology

The methodology developed for this study includes various steps such as –

1 Bathymetry data & Foundation base assimilation and categorization.
2 Exploring detrimental area: Fishing zones, shipping lines, environmental factors and marine protected areas.
3 Evaluation of potential area for power generation.
4 Calculation of theoretical number of turbines.
5 Wind speed data at hub height.
6 Power output per turbine.
7 Evaluating wind power potential

2.1 Bathymetry data and foundation base assimilation and categorization

Bathymetric data plays a crucial role in selecting the appropriate turbine foundation technology. Different technologies used vary with water depth as mentioned in table 1. Water depth in the offshore area has a direct impact on the design, construction and cost of turbine foundation technology. In the present study, bathymetry data has been retrieved from GEBCO (General Bathymetric Chart of the Oceans) with a spatial resolution of 30-arc second. Advanced jacket and floating foundations are under research & development. Based on bathymetry data, the study area has been processed and presented in figure1 (a). Monopiles are currently by far the most popular solution used worldwide, with 75% share, in comparison with only 5% for jacket/tripod options[10]. Within the area having water depth up to 30 m, maximum estimated available area is found near southern and western part of study area.

2.2 Exclusion zones

The 42 ports (one major, rest non-major) in Gujarat account for about a phenomenal 35% of the total traffic handled by all ports in India [11]. Suitable site for wind energy should not be located in a route junction or converging area of ships' routing or in any other way in the vicinity to maintain the risk to shipping at a minimum [12] A buffer of 2 km is provided on both side of shipping line considering full shipping constraint [13].

Table 1. Offshore wind turbine foundation technologies with appropriated water depth range and available area

Foundation Technology	Water depth range (m)	Available area (km^2)	Percentage area
Monopile	0-30	23935	10%
Jacket	30-50	7585	3%
Advanced Jacket	50-120	60152	25%
Floating	>120	149008	62%
Total		240680	100%

The environmental impacts of any offshore wind project need to be analyzed prior to its construction and during its operation. About 75% of the total marine fish landings of Gujarat are of comparatively low economic value. At present, there are 5 developed fishing harbors with good infrastructure, 18 developed landing centers and 117 other landing centers without any major infrastructure [11]. In order to calculate the distances at which fish can be affected by noise, models of sound propagation under water have been combined with knowledge of hearing and reaction of fish towards sound. Some studies considered noise caused by piling which is 80 kilometers away was being perceived by fishes such as herring and cod [14].
Considering Gujarat, regions highlighted with blue contours in figure 1(b) are important in this regard. Although wind energy is considered as environmental friendly, its development has caused a negative impact on the birds. These include displacement from existing habitats, risk of collisions with wind turbine blades, and migration barriers for birds. Thus to avoid displacement, certain areas should be restricted from installation of wind farms. Even though local bird species have adapted to wind farms, many vulnerable species are severely impacted through wind farms being established in their traditional feeding or breeding areas [14].

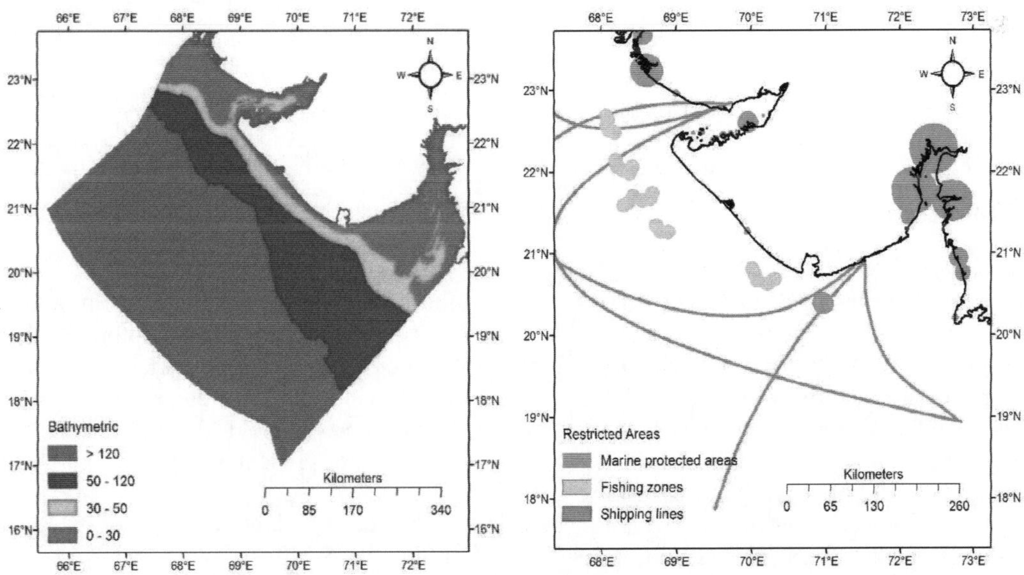

(a) **(b)**

Figure 1. (a) Classification of bathymetric data based on offshore wind turbine foundation technologies. (b) Exclusion areas for different utilities such as Marine protected area, fishing zones and shipping lines.

3 Calculation of offshore wind power potential

3.1 Analysis of wind data

Studies utilizing reanalysis wind dataset (ERA-Interim) to provide a summary of offshore wind energy potential in the exclusive economic zone of India indicate that reanalysis data can be considered as fit for estimation of available wind energy potential within such zones. ERA-Interim offers the greatest serial correlation with high-quality observations and greatest internal consistency over time [15]. For estimation of available wind energy potential further reanalysis data can be considered.

The main advantages of reanalysis data are the space and time homogeneity and the long series simulated. Of the many reanalysis data sets, studies indicate for the sample points chosen, ERA-Interim offers the greatest serial correlation with high-quality observations and greatest internal consistency over time [15]. Further reanalysis data can be considered as fit for estimation of available wind energy potential. The offshore wind resource is obtained based on monthly average reanalysis wind data-sets ERA-Interim made available by the European Centre for Medium-range Weather Forecasting (ECMWF)[16]. Datasets are obtained for 14 years from 2001 to 2014 at a spatial resolution of 0.125° (~12.5 km).

3.2 Wind energy analysis

The wind speed data is obtained at the height of 10 m above sea level, but we have to extrapolate it at a hub height of 80 m using equation (1).

$$V = V_{ref} \frac{\ln(z / z_o)}{\ln(z_{ref} / z_o)} \tag{1}$$

where, V is wind speed at hub height z, and z_{ref} is the height at which wind speed V_{ref} is measured. A surface roughness factor z_0 = 0.2 mm [17] is considered in the present study.

Power density is the amount of power generated at a particular wind speed per unit area swept by the wind turbine rotor (W/m^2). The mean wind speed and mean wind power density are estimated from equation (2) and equation (3).

$$\overline{V} = \frac{1}{N}\sum_{i=1}^{N} V_{\mathrm{i}} \tag{2}$$

$$\frac{P}{A} = \frac{1}{2}\rho \frac{\sum_{i=1}^{N} V_{\mathrm{i}}^{3}}{N} \tag{3}$$

Where V_i is the wind speed at observation i, ρ is wind density (1.225 kg/m^3) [19] and N is the number of observations.

3.3 Calculating wind power output

Low wind offshore regions are the regions with mean wind speed less than 7 m/s at 80 m above sea level and therefore the generated energy is limited by the swept area of the rotor[18,19]. The performance of the WT rotor is assessed by the rated rotor productivity (RRP), which is given as the ratio of rated capacity to swept area of WT. At high speed wind speeds, higher RRP metric WT achieves a higher Capacity Factor (CF). Accordingly, Siemens SWT 3.6-120 wind turbine having relatively large rotor and small generator ratio are chosen for the present study. The performance characteristics of the selected WT is described in Table 2 [20].

The net electric power generation (P) given in equation (4) can be calculated using information about the power curve taken from product catalogues [20] and of Siemens SWT 3.6-120 (SWT-3.6).

Capacity factor (CF) can be defined as the ratio of actual power generation to the rated power generation at a particular location, over a given period (T) as given in equation (5). In this study, CF is evaluated based on 14 years of wind data.

Table 2. Turbine characteristics for the sample wind turbine Siemens wind turbine

Operating data	SWT – 3.6 - 120
Rated Power (kW)	3600
Cut-in wind speed; V_{ci} (m/s)	3
Rated wind speed; V_r (m/s)	12
Cut-out wind speed; V_{CO}(m/s)	25
Rotor Diameter (m)	120
Swept Area (m^2)	11300
Hub Height (m)	site-specific

$$P = \begin{cases} 0 & V \le 3 \\ 1.138V_i^3 + 11.34V_i^2 - 2.526V_i - 98.18 & 3 < V \le 11 \\ 239.3V_i^3 - 8709V_i^2 + 105800V_i - 426200 & 11 < V \le 13 \\ 3600 & 13 < V \le 25 \end{cases} \tag{4}$$

$$CF = \frac{\text{Electric power generation } P(kWh/\,year)}{\text{Rated power}(kW) \times T(hours/\,year)} \tag{5}$$

3.4 Energy production

To calculate the power that could be produced by the offshore wind resource, first the number of turbines that could be fitted within the available area is determined. Subsequently, the wind regime power output of each turbine is analysed. The array spacing for the turbine is given by equation (6).

$$\text{Array Spacing} = (\text{rotor diameter})^2 \times \text{Downwind spacing factor} \times \text{Crosswind spacing factor} \tag{6}$$

Where, downwind space factor is 5 and crosswind space factor is 10.

$$\text{Number of turbines} = \frac{\text{Total available area}}{\text{Array spacing}} \tag{7}$$

Depending on array spacing, number of WTs that can be installed within the available area is estimated from the following equation (7). The nameplate wind capacity (i.e. total installed capacity) is found by multiplying the number of turbines by the nameplate capacity of a single turbine. The average output or Annual Energy Production (AEP) is more useful than installing nameplate capacity while comparing renewable energy resources such as offshore wind to traditional energy generation sources (i.e. coal, nuclear, natural gas). AEP is calculated by multiplying nameplate capacity with capacity factor as given in equation (8).

$$AEP = \text{CF} \times \text{nameplate capacity} \times \text{number of hours in the year} \tag{8}$$

The energy-produced regions of Gujarat area along with their corresponding depths were found and tabulated in Table 4. Given the available area, a plant with a potential of maximum 2580.46 TWh is feasible with SWT 3.6 on offshore conditions of Arabian Sea. The values of AEP represent the minimum assured power generation from the selected WT in the selected region.

4 Result and discussion

Foundation technology specific areas were calculated based on bathymetry data and wind speed. Available area in $km^2$and the percentage of the total area for various water depths and corresponding

IOP Conf. Series: Earth and Environmental Science **154** (2018) 012001 doi:10.1088/1755-1315/154/1/012001

foundation technology are as given in Table 1. Firstly, currently used monopile and jacket type foundation technology has 23935 km^2, which cover 10 % of the total area, and 7585 km^2, which covers 3% of the total area respectively. Secondly under R&D phase advanced jacket and the floating type foundation technology has 60152 km^2, which cover 25 % of the total area and 149008 km^2, which cover 62% of the total area respectively. From the Table 1 the net available area out of the total offshore area for the installation of wind turbine is obtained by eliminating the exclusion areas like shipping lines, fishing zones and protected marine areas. Hence, this net area must be used as efficient as possible for harnessing the offshore wind energy.

Spatial distribution of wind speed over study area at 80 m height is shown in figure 2(a). It depicts that the Kutch region state has highest wind speed along its coast as compared to the other coastal regions of the Gujarat. The Saurashtra region of the state has lower wind speed along its coastal region as compared to that of Kutch region but has greater value as compared to southern region of the state, which includes cities like Surat, Valsad, Vapi and many industrial areas.

Spatial distribution of capacity factor over study area at a 80 m height is shown in figure 2(b). It shows that the topmost region i.e. the Kutch region of the Gujarat state has the highest capacity factor as compared to the Saurashtra and the southern region of the state. The Saurashtra region has the capacity factor far less than that of the Kutch region but higher than that of the southern part of the state.

Results obtained from the available area, power potential and energy generation depth are shown in Table 3 below. Data like available area in km^2, number of wind turbines, nameplate capacity in (MW), average output (MWs), capacity factor and total annual generation (TWh/year) is obtained for various depths at which the turbine hub is placed. The value of all data as mentioned in the above statement is highest for the wind turbine at a depth greater than 120 m. The exclusion of area leads to a minor loss of potential of 225 TWh, with the conflict by designated shipping lines being the highest contributing factor. This exclusion amounts to energy potential of 121.7 TWh and an area subtraction of 11010 km^2. Thus, the heavy ship traffic significantly reduces the available area for offshore wind projects. Yet, the annual energy production is still substantial to provide feasibility of an offshore wind farm. Though the energy potential of the areas having water depth of 0-30 m & 30-50 m can substantiate a large amount of energy requirement for Gujarat, technological up gradation in floating turbines can help exploit wind potential at water depth greater than 50 m which is economically feasible.

Power potential and energy generation from the entire study area before and after subtracting exclusion areas like the shipping lines, fishing zones and marine protected areas is mentioned in table 3. Data like available area in km^2, number of wind turbines, nameplate capacity (MW), average output (MWs) and total power generation (TWh/year) is obtained for two different cases; firstly for the data for before considering restricted area and the second includes after subtracting the restricted areas. Here, the restricted area includes the shipping lines, fishing zones and marine protected areas. Table 4 summarizes that by considering the exclusion zones, the gross available area for offshore wind production decreases.

Table 3. Power potential and energy generation from the entire study area before and after subtracting exclusion areas.

	Available area (km^2)	Wind Turbines	Nameplate Capacity (MW)	Average Output (MW)	Total Annual Generation (TWh/year)
Before subtracting restricted area	240680	334278	1203401	320345	2806
After subtracting restricted area	218737	303801	1093684	294573	2580

Gujarat had energy requirement of approximately 94.9 GWh, while it produced annual surplus of 232MWh, and the country faced a deficit of 23 TWh [21]. Table 4 shows that within the 0-30 m depth range of offshore wind turbine allows the installation of nearly 24138 turbines, which on average could supply annual energy generation of 135 TWh, roughly 142% of the Gujarat Energy Requirement. And considering a depth of 0- 50 m, offshore wind has the potential to fulfil 522% of Gujarat's load. If fully developed out to greater depth, offshore wind power has the potential of total annual generation of 2580 TWh per year, which is twice the India's electric consumption of 1162 TWh/year.

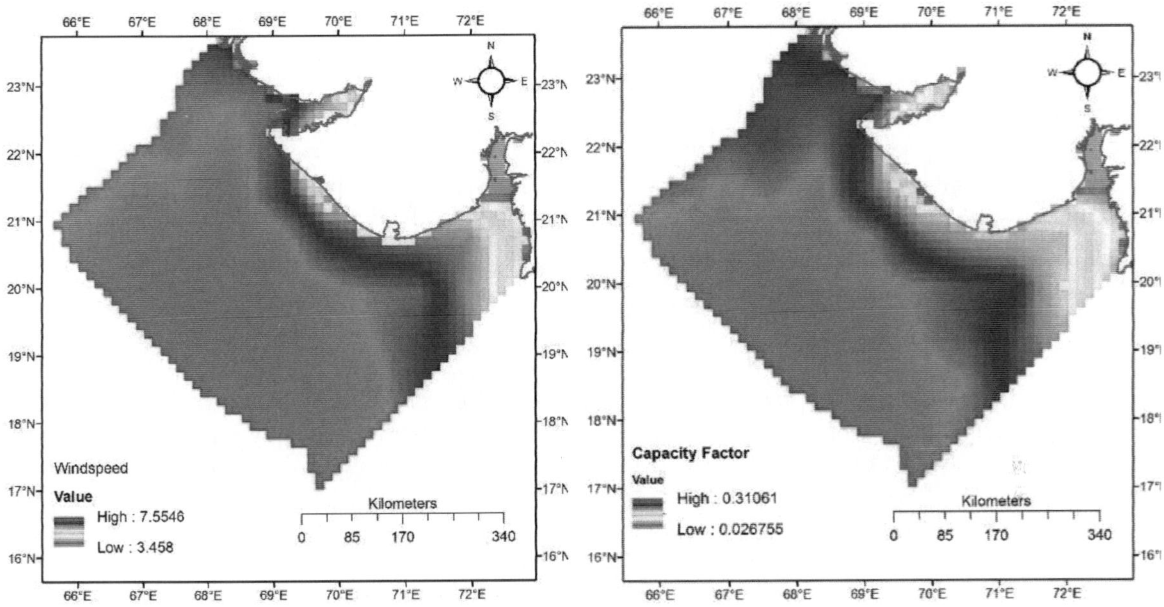

(a) **(b)**

Figure 2. (a) Spatial distribution of wind speed over study area at 80 m height. (b) Spatial distribution of capacity factor over study area at 80 m height

Table 4. Available area, power potential and annual energy generation corresponding to various depth.

Depth (m)	Available area (km^2)	Number of Wind Turbines	Nameplate Capacity in (MW)	Average Output (MW)	Capacity factor	Annual energy Generation (TWh/year)
0-30	17380	24138	86897	15496	0.18	135
30-50	5330	7403	26651	5659	0.21	49
50-120	52254	72575	261270	66044	0.25	578
>120	143773	199685	718866	207373	0.29	1816
Total	**218737**	**303801**	**1093684**	**294573**		**2580**

5 Conclusion

The potential annual energy production using SWT 3.6-120 is found to be 2580 TWh, more than twice the annual energy demand of India in 2016.

The study used reanalysis and bathymetric data show that wind potential in 0-30m depth has potential of 135.8 TWh per year which is feasible with current technologies.

Due considerations for marine protection zones, fishing zones, shipping lines and migratory bird movements have been accorded and corresponding area expulsion is done. The available area is found to be 90.91% of total area. Shipping lines corresponding to highest reduction with to 4.57% of total available area. The resulting reduction of 8% in wind resource potential is not significant to adversely affect the feasibility of wind farm project.

The study methodology presented a robust wind farm assessment with a consideration of environmental, shipping transportation and fishing zones for offshore areas on Gujarat. A careful consideration of overall impact of such offshore wind projects would be helpful for policy formulation.

6 References

[1] GWEC 2016 Global Wind Report 2015 | Gwec *Wind energy Technol.* 75
[2] E. Kondili and J. K. Kaldellis. 2012 *Environmental-social benefits/impacts of wind power* vol 2 (Elsevier Ltd.)
[3] Anon Gujarat may get India's first offshore wind power project
[4] S. B. Kore and S. L. A A G D A S S 2016 Feasibility of Offshore Wind Farm in India 995–8
[5] M. Burlando, A. Podest, L. Villa, C. F. Ratto and G. Cassulo. 2009 Preliminary estimate of the large- scale wind energy resource with few measurements available: The case of Montenegro
[6] B. Oller. 2010 Continuous spatial modelling to analyse planning and economic consequences of offshore wind energy
[7] A. Dhanju, P. Whitaker and W. Kempton 2008 Assessing offshore wind resources: An accessible methodology *Renew. Energy* **33** 55–64
[8] W. Musial and S. Butterfield. 2004 Future for Offshore Wind Energy in the United States: Preprint
[9] W. Musial and B. Ram 2010 Large-Scale Offshore Wind Power in the United States: Assessment of Opportunities and Barriers, NREL (National Renewable Energy Laboratory)
[10] M. Arshad and B. C. O'Kelly. 2013 Offshore wind-turbine structures: a review *Proc. Inst. Civ. Eng.-Energy* **166** 139–52
[11] Dr. Chandanathil P. Geevan and Dr. Arun Mani Dixit 2012 State of Environment Report on Coastal and Marine Environment Gujarat State
[12] M. Wg. Interaction between offshore wind farms and maritime navigation Terms of Reference
[13] L. Bergstrom, L. Kautsky, T. Malm, H. Ohlsson, M. Wahlberg, R. Rosenberg, N. A. Å. Capetillo, L. Bergström, L. Kautsky, T. Malm, H. Ohlsson, M. Wahlberg, R. Rosenberg and N. A. Å. Capetillo. 2012 *The effects of wind power on marine life. A synthesis*
[14] S. Wang, S. Wang and P. Smith. 2015 Ecological impacts of wind farms on birds: Questions, hypotheses, and research needs *Renew. Sustain. Energy Rev.* **44** 599–607
[15] M. C. Brower, C. T. Officer and M. S. Barton. 2013 A study of wind speed variability using Global Reanalysis data 12
[16] D. P. Dee, S. M. Uppala, A. J. Simmons, P. Berrisford, P. Poli, S. Kobayashi, U. Andrae, M. A. Balmaseda, G. Balsamo, P. Bauer, P. Bechtold, A. C. M. Beljaars, van de L. Berg, J. Bidlot, N. Bormann, C. Delsol, R. Dragani, M. Fuentes, A. J. Geer, L. Haimberger, S. B. Healy, H. Hersbach, E. V. Hólm, L. Isaksen, P. Kållberg, M. Köhler, M. Matricardi, A. P. McNally, B. M. Monge-Sanz, Morcrette J-J, Park B-K, C. Peubey, P. de Rosnay, C. Tavolato, Thépaut J-N and F. Vitart. 2011 The ERA-Interim reanalysis: configuration and performance of the data assimilation system *Q. J. R. Meteorol. Soc.* **137** 553–97
[17] F. Pimenta, W. Kempton and R. Garvine 2008 Combining meteorological stations and satellite data to evaluate the offshore wind power resource of Southeastern Brazil *Renew. Energy* **33** 2375–87
[18] FOWIND-Gujarat 2015 Pre-feasibility study for offshore wind farm development in Gujarat
[19] G. Nagababu, S. S. Kachhwaha, N. K. Naidu, V. Savsani. Application of reanalysis data to estimate offshore wind potential in EEZ of India based on marine ecosystem considerations. Energy. 2017;118.
[20] Siemens Wind Power 2009 Wind Turbine SWT-3.6-120 Technical Specifications 3
[21] Anon Central Electricity Authority. Load generation balance report 2015-16. New Delhi. 2015

IOP Conf. Series: Earth and Environmental Science **154** (2018) 012002 doi:10.1088/1755-1315/154/1/012002

Climate conditions of the "El Niño" phenomenon for a hydro-eolic complementarity project in peru

Leonardo Castillo N[1], Arturo Ortega M[2] and Jaime E Luyo[3]

[1]Civil Engineering, National University of Engineering, Lima-Peru (e-mail: leonardocastillo@uni.edu.pe).
[2]Mechanical Engineering, Pontificia Universidad Catolica del Peru, Lima-Peru (e-mail: aortegam@pucp.edu.pe).
[3]Mechanical and Electrical Engineer, National University of Engineering, Lima, Peru (e-mail: jluyok@uni.edu.pe).

Abstract. Northern Peru is threatened by the consequences of a natural phenomenon called "El Niño", mainly during the months of December to April. In the summer of 2017, this event reported strong climatic variations with intense rains, increasing the water levels of the Chira and Piura rivers, filling the Poechos reservoir, together with flooding and mudding. However, from an energetic perspective, these climatic alterations have a strong potential to increase the availability of the wind and hydro renewable energies in northern Peru. This work performs an evaluation of the hydro-eolic complementarity as part of the sustainability of energy systems. The study includes evaluation of historical records of wind velocity and water flow rates. It then evaluates correlation, analysis, and estimates the hydro and wind energy potentials generated by this phenomenon. The implications of the "El Niño" phenomenon are mostly negative. Nonetheless, it is possible to take advantage of higher wind and water flow rates with a hybrid energy system. The results obtained show a high degree of complementarity both normal and "El Niño" phenomenon condition in northern Peru.

1. Introduction

Originally the term El Niño was used by fishermen in Paita (city of northern Peru in Piura) to describe the warming of the sea in the North Pacific Peruvian (from southern Ecuador) and rains that were presented at Christmas time and for periods of several months (Carrillo, 1892) [1]. Bjerknes (1966, 1969) [2], [3] documents consistent heating anomalies of the Peruvian sea on the border with Ecuador and relates this behavior to both the warming of the tropical Pacific Sea and to the planetary scale changes of the tropical atmosphere, the "South Oscillation ". Hence the term ENOS (El Niño oscillation of the south) is derived. The warm phase of ENOS is called the El Niño while the cold phase is called La Niña (Wang et al., 2012) [4].

ONI (Oceanic El Niño index) is an estimated ENSO index or measure that determines periods of warming over the years along the Pacific coast. These values are reported on the National Oceanic and Atmosferic Administration (NOAA) Web site. It is evidenced, that the maximum values ONI reached indices of up to 2.2 and were given in the years 1982/1983 and 1997/1998.

During the last few decades, the implications that El Niño leaves in the north of Peru are mostly negative, but it is possible to obtain benefits for the energy sector. This paper analyses time series and introduces the concept of "complementarity" which includes two extraordinary events 1982/1983 and 1997/98.

IOP Conf. Series: Earth and Environmental Science **154** (2018) 012002 doi:10.1088/1755-1315/154/1/012002

Considering the technical challenges and the current state of the energy market, solutions must be proposed for the take-off of the capacities of renewable energies. The complementarity between different intermittent sources seeks an optimal balance between the total amount of energy produced and its stability over time. This complementarity reduces the variability of energy production, combining systems that have of minimum and maximum energy generation during different periods of time.
Based on new environmental, social and economic policies, the integration of renewable energies is a focus of attention in many countries. As a result, these systems become more attractive when we identify complementarity with other sources of generation. Therefore, the development of renewable energy systems that seek to "empower" intermittent power production is a key point in the sustainability of future energy systems.

2. Review of data

2.1.Historical records

For the development of work in the region of Piura will analyze a hydrometric station in the River Chira and two weather stations. The measurement stations to be analyzed are shown in Table 1:

Table 1. Locations of stations

Station	Latitude	Longitude	Altitude	Years
Pte. Sullana - Q(m3/s)	4°53'29"	80°41'28"	36	1972-2015
San Miguel - Vv (m/s)	5°18'14"	80°40'50"	20	1975-2015
Miraflores - Vv (m/s)	5°10'0"	80°36'0"	30	1995-2015

We note that records 2016 and 2017 years are not available within the official institution of Peru.

2.2.Final stage

The Chira River comes from a watershed regulated by the Poechos Reservoir. Upstream of the reservoir, the contributing flow is a border with Ecuador. For our case, measurements will be analyzed in the section downstream of the reservoir, near the city of Piura. The average monthly flow rate for the entire analyzed data is 99.57 m3/s and the maximum value corresponds to the month April 1983 with 1645.97 m3/s. Figure 1 shows the monthly flow rate, where the maximum values for the summer periods of 1982/1983 and 1997/1998 stand out. Figure 2 shows the monthly average values and the flows to a persistence of 95% at monthly level.

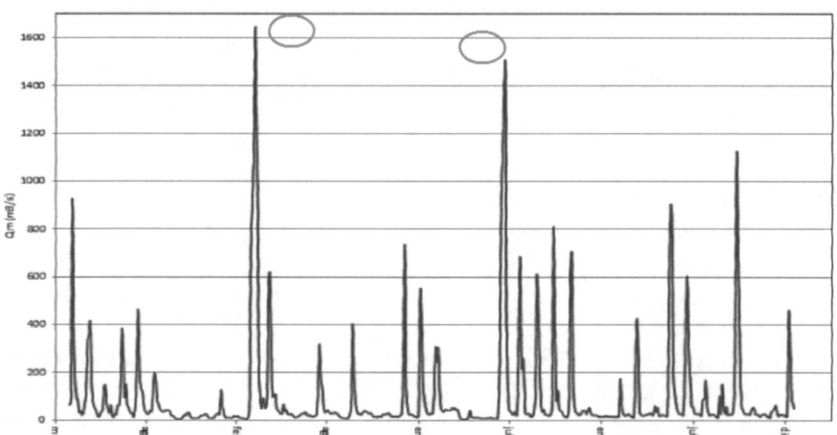

Figure 1. Monthly flow regime of the Chira river – Sullana bridge

IOP Conf. Series: Earth and Environmental Science **154** (2018) 012002 doi:10.1088/1755-1315/154/1/012002

2.3. Wind speed regimen

The San Miguel station has an average wind velocity of 4.5 m/s and a maximum value of 9.2 m/s registered in November 1986. The Miraflores station has an average wind velocity of 2.5 m/s and a maximum value of 4.8 m/s registered in September 1995.

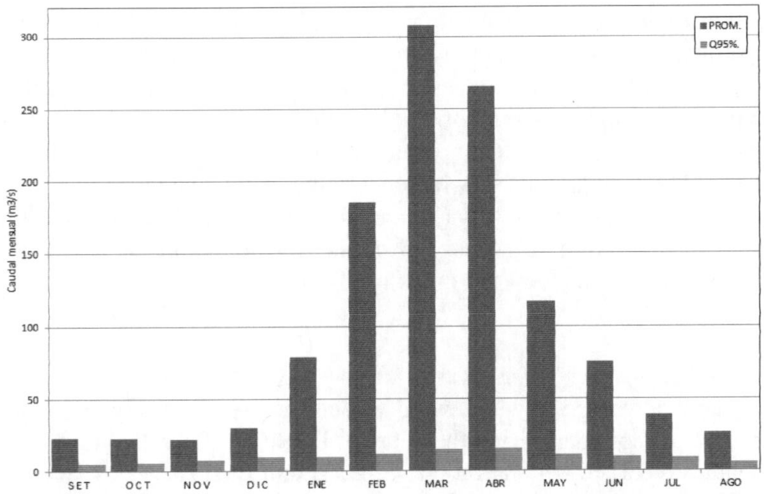

Figure 2. Average monthly flows of the Chira river on the Sullana bridge

To be able to obtain a more representative value of the wind speed it is necessary to describe it as a function of the distribution of known probability. The Weibull type distribution is best suited for monthly wind speed data (E. Akpinar, 2005) [5]. In the present work, the model HYFRAN (Hydrological Frequency Analysis) of the National Institute of Scientific Research Water Land and Environment (INRS-ETE) of the University of Québec was applied for the adjustment of the distribution and obtaining of the parameters. Figure 3 shows the average speed regime obtained from the Weibull distribution month to month.

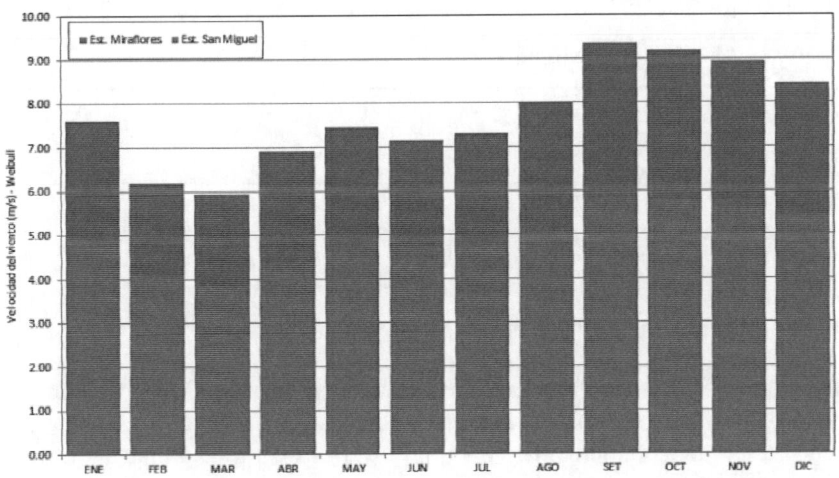

Figure 3. Average monthly speeds (m/s)

3. Methodology

As a first assessment, the complementarity of the flow rate regime with the wind speeds at the monthly level will be analysed. A simple method of execution and interpretation is by determining the Pearson correlation coefficient (δ). This coefficient varying between -1 and +1 is a value that measures the degree of covariance between different linearly related variables.

In the case of "complementarity", what we are looking for is to know if the relationship is perfect negative: when exactly to the extent that increases one variable decreases the other. Once verified the possibility of complementing energy projects, will estimate the hydroelectric and wind potential under an assumed stage. There will be applied the concept of an index of complementarity (Beluco et al., 2008). [6]

Beluco et al. [6], proposes an index to evaluate how two different types of energy are complemented. The proposed complementarity index considers three types of complementarity:

- Complementarity in time (Kt): Evaluates the time lag and compares in which days of the year happen the maximum and minimum of water a wind generation.
- Complementarity in energy (Ke): Compares the total amount of energy that water generation can produce and the total amount of wind energy that can be generated.
- Complementarity in energy availability (Ka): evaluates the amplitude of the variations between

The maximum level and minimum energy level available, for both water generation and wind. Based on these three concepts, the complementarity index is K = Kt * Ke * Ka.

4. Results

4.1. Complementarity with registered average values

From the recorded data, the average flow and wind velocity values are analysed. Within the scenarios, it is considered the effect of not considering the years of El Niño (1972/1973 and 1982/1983). Table 2 shows the average values recorded per station (QCC, ESM and EMI scenarios) and the average values without El Niño (QCS scenario).

Table 2. Average values of data

QCC: Flows (m3/s) with El Niño – Pte. Sullana											
Ene	Feb	Mar	Abr	May	Jun	Jul	Ago	Set	Oct	Nov	Dic
79.05	185.43	307.66	264.98	116.73	75.15	39.33	27.15	23.64	22.79	22.51	30.39
QCS: Flows (m3/s) without Niño – Pte. Sullana											
Ene	Feb	Mar	Abr	May	Jun	Jul	Ago	Set	Oct	Nov	Dic
42.76	144.66	260.93	201.25	81.28	52.19	34.40	25.43	24.46	23.52	23.18	25.37
ESM Wind velocity (m/s) – Est. San Miguel											
Ene	Feb	Mar	Abr	May	Jun	Jul	Ago	Set	Oct	Nov	Dic
4.46	3.79	3.57	3.98	4.39	4.27	4.37	4.62	5.27	5.26	5.27	4.95
EMI: Wind velocity (m/s) – Est. Miraflores											
Ene	Feb	Mar	Abr	May	Jun	Jul	Ago	Set	Oct	Nov	Dic
2.44	1.84	1.82	2.27	2.35	2.19	2.24	2.69	3.25	3.12	2.90	2.71

Table 2 shows that the average flow values for the Chira River vary significantly in the summer period if we do not consider the years of El Niño, reaching a decrease of up to 46% in January.

Table 3 gives the results of complementarity with values close to -1, which indicates a contrast of max/min between variables.

4.2. Complementarity with insurable values

The same complementarity assessment is done with insurable values. For the case of flows, the flow is available for a persistence of 95% for small hydroelectric plants in the past. For the case of wind speeds, the most probable value obtained from the parameter is worked and adjusted with the distribution of Weibull. More detail See Table 4.

Table 3. Complementarity results to average values – Pearson Coefficient

Variables	δ
QCC-ESM	-0.85
QCC-EMI	-0.74
QCS-ESM	-0.82
QCS-EMI	-0.72

In Table 5 there happen the results of the coefficient of Pearson with values near to -1, what indicates a complement between variables at the level of available values. This complementarity is evidenced in Figure 4.

Table 4. Insurable average values

QCC: Flows (m3/s) with El Niño – Pte. Sullana											
Ene	Feb	Mar	Abr	May	Jun	Jul	Ago	Set	Oct	Nov	Dic
10.12	11.91	15.36	16.39	11.22	9.92	9.33	6.34	5.68	5.94	8.01	9.76
QCS: Flows (m3/s) without Niño – Pte. Sullana											
Ene	Feb	Mar	Abr	May	Jun	Jul	Ago	Set	Oct	Nov	Dic
9.80	11.91	15.35	16.267	11.04	9.89	9.315	6.13	6.08	6.44	23.18	25.37
ESM Wind velocity (m/s) – Est. San Miguel											
Ene	Feb	Mar	Abr	May	Jun	Jul	Ago	Set	Oct	Nov	Dic
4.89	4.11	3.87	4.37	4.84	4.71	4.83	5.05	5.82	5.79	5.79	5.44
EMI: Wind velocity (m/s) – Est. Miraflores											
Ene	Feb	Mar	Abr	May	Jun	Jul	Ago	Set	Oct	Nov	Dic
2.72	2.07	2.06	2.54	2.61	2.44	2.47	2.95	3.52	3.40	3.14	2.98

Table 5. Complementarity results to insurable values-Pearson Coefficient

Variables	δ
QCC-ESM	-0.83
QCC-EMI	-0.78
QCS-ESM	-0.81
QCS-EMI	-0.77

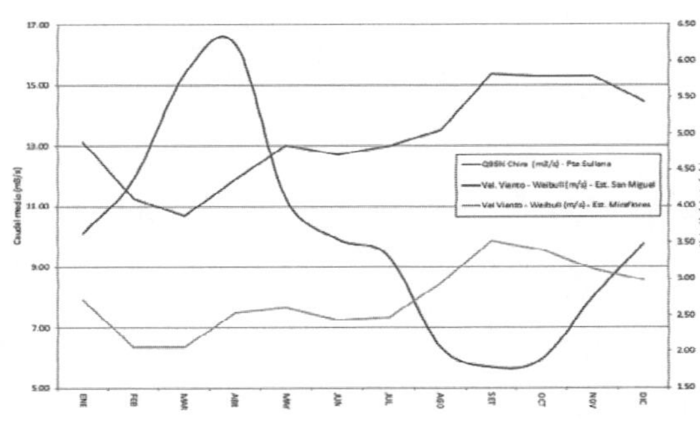

Figure 4. Comparison of flow rate regime with wind speed at monthly level

4.3. Complementarity by the index of A. Beluco et. Al.

$$Ph = \gamma * Q_{95\%} * H * n$$
$$Pe = Cp * 0.5 * \rho * A * V^3$$

Table 6. Results of energy y generation

Month	Ph (MW)	Pe (MW)	Ph (MW)	Pe (MWh)
Jan	2.4	0.33	1773.4	246.8
Feb	2.8	0.20	1884.3	132.3
Mar	3.6	0.16	2690.6	122.3
Apr	3.9	0.24	2778.4	170.5
May	2.6	0.32	1965.4	239.3
Jun	2.3	0.30	1681.6	213.4
Jul	2.2	0.32	1634.3	237.8
Aug	1.5	0.37	1110.6	271.8
Sep	1.3	0.56	963.2	402.7
Oct	1.4	0.55	1040.7	109.7
Nov	1.9	0.55	1358.5	396.5
Dec	2.3	0.46	1709.6	339.8

Table 7. Complementarity Index

Equation	Parameters	Results
$K_t = \frac{\lvert d_h - d_s \rvert}{\sqrt{\lvert D_h - d_h \rvert \lvert D_s - d_s \rvert}}$	d_h: day of lower availability of hydropower d_s: day of lower availability of wind energy D_h: day of greater availability of hydric energy D_s: day of greater availability of wind energy	d_h=258 d_s=74 D_h=105 D_s=288 K_t=1.017
$K_e = 1 - \sqrt{\left(\frac{E_h - E_s}{E_h + E_s}\right)^2}$	E_h: Water energy per year E_s: Wind energy per year	E_h=20.59 GWh E_s=3.18 GWh K_t=0.268
$K_t = \begin{cases} \left[1 - \frac{(\delta_h - \delta_s)^2}{(1-\delta_s)^2}\right] \\ \left[\frac{(1-\delta_s)^2}{(1-\delta_s)^2 + (\delta_h - \delta_s)^2}\right] \end{cases}$ $Para\ \delta_h \leq \delta_s$ $Para\ \delta_s \leq \delta_h$	$\delta_s = 1 + \frac{E_{s,max} - E_{s,min}}{E_{s,prom}}$ $\delta_h = 1 + \frac{E_{h,max} - E_{h,min}}{E_{h,prom}}$	Es,max=409.7 MWh Es,min=122.3 MWh Es,prom=265.2MWh Eh,max=2778.4 MWh Eh,min=963.2 MWh Eh,prom=1715.9MWh K_t= 0.999
$K = K_t K_e K_a$	Complementarity Index	K=0.272

In the order to determine the water and wind generation, we are based on the following technical approach: the reach of the Chira River where the flows are analyzed, from the downstream point of the reservoir to the Sullana Bridge, there is a runway of 34 Km with a gap (H) of approximately 30 m. We'll assume a hydroelectric project with a global efficiency (n) of 80%.
In the case of the wind generation, we will assume a wind turbine Vestas V100-1.8MW with an area swept rotor (A) of 7 850 m2. In addition to considering the factor of Beltz (Cp) 59 % for the useful potency.
Table 6 shows the results of the calculations by applying the power equations.
Finally, we will obtain the parameters of the complementarity index to validate that the flow behavior (including the data for El Niño) can be complemented by a wind project in the region. Table 7 it shows the equations and results of the procedure given by Beluco et al. [6]
The results shown in Table 7 are beneficial because there is a good complementarity in time (Kt) and energy availability (ka). Regarding the complementarity of energy (Ke), can be managed and adapted to better design conditions that allow optimization of the system.
The flow and wind conditions in the northern region of Peru are suitable for implementing a complementarity project assuming the entire historical record including the periods of El Niño.

5. Conclusions

The average monthly flow of the Chira River is 99.57 m3/s and the maximum value during the last few decades correspond to the month April 1983 with 1645.97 m3/s. A phenomenon of El Niño may imply a multiplicative factor in the flow of approximately 15 times.
The San Miguel station has an average wind velocity of 4.5 m/s and a maximum value 9.2 m/s registered in November 1986. The Miraflores station has an average wind velocity of 2.5 m/s and a maximum value of 4.8 m/s registered in September 1995. The area for the San Miguel area has higher speeds for its use.
The Pearson coefficient indicates the high complementarity between the variables of flow and wind velocity at monthly level, with a value of -0.81 close to the unit.
The complementarity index of Beluco et al., can be optimized based on established design conditions. In this paper, it has obtained a crude hydroelectric power under a flow to persistence of 95% and for an efficiency of 80%. In the case of the wind project the calculation only considers the potency a typical wind turbine.
It is possible to take advantage of the generation of energy in the northern area of Peru, trough hydro-eolic complementarity projects. A next step is to evaluate the technical feasibility of a hydroelectric project that can take advantage of the increase in flow in events of El Niño.
A next step is to proceed with the measurements and monitoring in order to be able to validate the complementarity and process on based on statistics and modelling.

6. References

[1] KPINAR E Kavak, S. Akpinar, "A statistical analysis of wind speed data used in installation of wind energy conversion systems", *Energy Conversion and Management, vol.* **46**, *p.* **515–532, 2005**.
[2] BELUCO A P Kroeff de Souza, A. Krenzinger, "A dimensionless index evaluating the time complementarity between solar and hydraulic energies", *Renewable Energy, vol.* **33**, *p.* **2157–2165, 2008**.
[3] J. BJERKNES, "A possible response of the atmospheric Hadley circulation to equatorial anomalies of ocean temperature". *Tellus Series a-Dynamic Meteorology and Oceanography, v.* **18**, p. **820-829, 1966**.
[4] J. BJERKNES, "Atmospheric teleconnections from the equatorial Pacific". Monthly Weather *Review, v.* **97,** *p.* **163-172, 1969**.
[5] C. CARRILLO, "Hidrografía oceánica: Las corrientes oceánicas y estudios de la Corriente Peruana o de Humbolt". Boletin de la Sociedad *Geográfica de Lima, v.* **2**, *p.* **72-110,1892**.

[6] C. WANG; C. DESER, YU, J.-Y; P. DINEZIO, A. CLEMENT, "El Niño-Southern Oscillation (ENSO): *A review*". In “Coral Reefs of the Eastern Pacific”, P Glymn, D Manzello, and I Enochs, *Eds., Springer Science Publisher*, **2012**.

Promotion of renewable energy in some MENA region countries

K. Abdeladim, S. Bouchakour, A. Hadj Arab, S. Ould Amrouche, N. Yassaa

Centre de Développement des Energies Renouvelables, CDER, BP 62, Route de l'Obsevatoire, Bouzaréah, 16340, Algiers, Algeria
E-mail: k.abdeladim@cder.dz

Abstract. In recent years Middle East and North African (MENA) countries, are showing efforts about the integration of renewable electricity into their power markets. Indeed, installations were already achieved and renewable energy programs were launched. The Algerian program remains one of the most ambitious with its installation capacity up to 22GW of power generating to be installed by 2030. More than 60 % of the total capacity is planned to be solar photovoltaic (PV). Like Algeria, Morocco has integrated development project with a target to develop by 2020 a 2000 MW capacity of electricity production from solar energy. The Tunisian government has launched its first phase of the renewable power generation program, with an objective to install 1,000 MW of renewable power capacity over the 2017-2020 periods, where 650 MW of the total capacity is planned to be solar and 350 MW wind. One of the leading Arab country in wind energy, these recent years is Egypt, with its more than 700 megawatt of operational power generation plants and has launched significant projects development in solar energy. Regarding Jordan, the government has taken different steps in this field of energy with a Strategy plan 2007-2020, by implementing a large scale of projects on renewable energy sources, with an objective to cover 10% of the country's energy supply, from renewable sources by the year 2020. Concerning Lebanon, the country is looking to attain an integration of 12 % by 2020.

1. Introduction

The MENA region heavily dependents on non-renewable, fossil energy sources. High population growth and energy-intensive consumption in the region will cause energy needs to surge by 6 to 7% per year in the future, if consumption patterns stay the same. The energy needs are estimated to double by the year 2020 [1].

In order to face the increasing demand for energy, many countries in the MENA region started to move smoothly the energy mix; this will lead to the reduction of greenhouse gas emission. The subsidization of fossil energy sources in the region is putting a tremendous strain on government budgets. The subsidization regime is not contributing to socially just support for the weakest private households.

Nowadays, renewable energy plays a great role, is being as a potential means of industrial diversification, new value-chain and employment activities, technology transfer, and improved environmental paths. Even though wind has become the largest renewable energy source after hydro, solar power generation has seen higher growth in recent years, through photovoltaic (PV), eventually due to the price drop. Like PV, we saw a commissioning of large concentrating solar power (CSP) plants in Algeria, Egypt, Iran, and Morocco. Furthermore, solar water heating (SWH) systems are also a key element in the deployment of renewables in the region.

2. Situation in MENA countries

Please follow these instructions as carefully as possible so all articles within a conference have the same style to the title page. This paragraph follows a section title so it should not be indented.

2.1. Case of algeria

The availability of large amount of solar potential in Algeria means it would be possible to consider solar energy as a potential source of energy for different applications as individual modules or solar photovoltaic systems.

In 2011, the Algerian government has launched an ambitious programme dedicated for renewable energies plans to produce 22,000 MW of electricity from renewable sources, including solar and wind, for the local market, with more than 10,000 MW export for the next 20 years. This corresponds to 40% of global electricity production by 2030 and double the current capacity of the national park of electricity; this programme was updated recently, where the capacity of the photovoltaic solar plant was increased to 13.575 MW.

The total capacities for different considered technologies are given in table 1 [2].

Table 1. Total capacity to be installed.

Technology	Capacity
Phototovoltaics	13575 MW
Wind	5010 MW
Thermal	2000 MW
Biomass	1000 MW
Cogeneration	400 MW
Geothermal	15 MW

2.2. Installed capacity.

Different installations were achieved so far, recently a call tender was released for the installation of 4 000 MW of PV, below are given some examples of installed plants, SKTM (Shariket Kahraba wa Taket Moutadjadida) are responsible for these projects:

- 23 PV plants were installed across the country representing a capacity of more than 343 MW.
- Pilot PV plan of 1.1 MW was installed in the region of Ghardaia
- Wind farm of 10 MW in the region of Tindouf.
- A hybrid solar-gas plan with a capacity of 150 MW with 25 MW solar in the region of Hassi R'mel. Concentration solar collector is provided by parabolic trough.

Energy Efficiency Policy. In February 2015, the Algerian government announced a new national programme on energy efficiency (EE) for the years 2015 to 2030. The EE programme mainly targets three sectors: the building sector, transport and industry [3].

In the building sector, more than 30 million toe are supposed to be saved by 2030. This will be achieved through innovative technologies and thermal insulation of constructions. In addition, economic lamps and solar water heaters will contribute. In the transport sector, the target is to save more than 15 million toe. The government aims to use fuels, which are widely available and least polluting, like LNG and natural gas, in order to reduce the use of gasoline. The industry sector is supposed to contribute 34 million toe by 2030.

2.3. Case of morocco
Morocco is one of the leading potential markets for renewable energy across the MENA region. The country is targeting sufficient renewable energy to account for 42% of its energy generation mix by 2020. In 2009, the government has planned to develop 2,000MW of solar capacity across five sites. It also plans to bring online 2,000MW of wind capacity by 2020 [4]. Morocco has a potential to export renewable energy to southern Europe.

2.3.1. Wind. Some large projects are in the pipeline, includ¬ing GDF Suez and Nareva Holding's 300MW Tarfaya wind farm, which started construction in June 2013, and Nareva Holding's 50MW Haouma and 50MW Foum El Oued wind farms, which are due to begin con¬struction imminently.

2.3.2. *Solar.* Solar energy in Morocco was playing a minor role in the Moroccan energy mix, mostly in the rural electrification programme, where photovoltaic (PV) power was used to supply isolated villages. There have been substantial changes in this field since the government's solar power plan, entitled the Moroccan Solar Power Programme, was launched back in 2009. The programme has an objective to increase installed capacity in terms of CSP and PV generation to at least 2,000 MW by 2020. In 2015, two sites, located in Midelt and Tata, were added to the original five. The development of these two sites has been carried forward, which will allow Morocco to exceed the 2,000 MW mark by 2020.

Table 2, shows the seven sites where major projects have been carried out or are scheduled for installation.

Table 2. Total capacity (have been carried out or to be installed).

Site/power plant	Power in MW	Technology
Ouarzazate	**Total: 580**	
NOOR I	160	CSP
NOOR II	200	CSP
NOOR III	150	CSP (solar tower)
NOOR IV	70	PV

2.4. Case of tunisia
Tunisia is an energy-dependent country with modest oil and gas reserves. Around 97 percent of the total energy comes from natural gas and oil, while renewables contribute at a level of 3% of the energy mix. The installed electricity capacity at the end of 2015 was 5,695 MW, which is expected to sharply increase to 7,500 MW by 2021 to meet the increasing power demands. Meanwhile, Tunisia is building additional conventional power plants and developing its solar and wind capacities to withstand its economic growth [5].

2.4.1. *Wind.* Tunisia holds an important wind resource; In fact, wind power represents the leading source of renewable energy in Tunisia. Since 2008, wind energy is considered a principal the energy transition of Tunisia with a growth of the production up to 245 MW of power installed in 2016. Two main wind farms have been implemented until now: Sidi-Daoud and Bizerte.

2.4.2. *Solar.* Back in 2005, the government has launched a Solar Program called PROSOL, is a joint initiative of UNEP, Tunisian National Agency for Energy Conservation, state-utility STEG and

Italian Ministry for Environment, Land and Sea. The program has an objective to encourage the enlargement of the solar energy sector via financial and fiscal support. PROSOL includes a loan mechanism for domestic customers to purchase Solar Water Heaters and a capital cost subsidy provided by the Tunisian government of 20% of system costs. The major benefits of PROSOL are:

- More than 50,000 Tunisian families get their hot water from the sun based on loans
- Generation of employment opportunities in the form of technology suppliers and installation companies.
- Reduced dependence on imported energy carriers
- Reduction of GHGs emissions.

The Tunisian Solar Plan includes 40 projects with an objective to promote solar thermal and photovoltaic energies, wind energy, as well as energy efficiency measures. Besides, the plan includes the ELMED project; a 400KV submarine cable interconnecting Tunisia and Italy.
The first large scale solar power plant of a 10MW capacity, co-financed by KfW and NIF (Neighbourhood Investment Facility) and implemented by STEG, is due 2018 in Tozeur.
Concerning CSP technology, the government has launched a program called "TuNur". The program is considered as Tunisia's most ambitious renewable energy project yet. The project includes of a 2,250 MW solar CSP plant in Sahara desert and a 2 GW HVDC (High-Voltage Direct Current) submarine cable from Tunisia to Italy. The project itself will generate 2.5GW of electricity on 100km2 of desert in South West Tunisia by 2018.

2.5. Case of egypt

Egypt holds a rich renewable resources, which could be an interesting solution in Egypt. As a matter in fact, the government has given an importance to the use of renewable in order to meet the increasing demand in energy supplies. PV and wind technologies are the main options for electricity generation mix in Egypt Among different renewable energy technologies,. Unfortunately, other options such as solar water heaters, biogas and biomass technologies are not yet given enough attention, except for some donor-funded activities and efforts of NGOs that try to build some momentum on pilot and local scales, but not yet under a national deployment program [6].
The renewable equipment market is potentially worth billions of dollars. By the year 2022, the Egyptian government has an intention to reach 20 percent of generated electricity from renewable sources (Wind 12 %, Hydropower 5.8 %, and Solar 2.2 %). The solar energy plan has an objective to set up 3.5 GW by 2027 including 2.8 GW of PV and 700 MW of CSP. Besides, there is plan to produce 7.2 GW (12 percent of generated electricity) from wind by 2022, where private sector will be involved (67 % of the plan). Over the next three to five years the Ministry of Electricity and Renewable Energy has an aim to increase 51.3 GW to current installed capacity.
In June 2014, the total installed capacity reached about 32 GW (28.8 thermal, 2.8 hydro & 0.55 wind).

2.5.1. *Wind.* Egypt holds a huge wind potential along the Suez Gulf with an average wind speed of 10.5 m/sec. The Government has an objective to reach 12 % of the generated electricity (6.8 GW) through generated wind energy by 2022.

Since 2001, a series of large-scale wind farms have been implemented with total capacity of 550 MW in cooperation with Germany (KFW), Denmark (DANIDA), Spain, and Japan (JICA). Implementation of the Spanish project in Jebel El Ziet took place in 2013, which clarifies the huge increase in Egypt imports for renewables, which were mainly wind powered generators imported from Spain.

2.5.2. *Solar.* In 2011, the first Solar Thermal Power Plant at Kuraymat was established, with a total installed capacity of 140MW where, 20MW are based on parabolic trough technology integrated with combined cycle power plant using natural gas. A 10 MW power plant also has been operating in Siwa since March 2015.

2.6. Case of jordan

Jordan started the promotion of renewable energy since early 1970s. The first wind power pilot project was installed back in 1988. In recent years, the government gave a great importance to the generation of power from renewable with many initiatives. By 2020, the renewable energy integration should attain 10% [7].

2.6.1. *Wind.* Jordan holds an important wind energy potential. Actually, the government has prepared an ambitious program in wind energy development, where about 1200MW of wind turbines are to be installed by the year 2020.

2.6.2. *Solar.* Concerning solar energy potential in Jordan, the resource is great, where solar radiation ranges between 5 and 7KWh/m2. Concerning the total installed capacity, it still to be improved, where only 1 MW of solar installation is available (photovoltaic kits including water pumping in rural and remote villages). Meanwhile, the good news is that 15% of all households are equipped with solar water heating systems. A goal the increase the number up to 30% in the future (by 2020). Concentrated Solar Power (CSP), has its share in the program, in fact the government is aiming to set up the first (CSP) demonstration project in the short to medium term. In addition, a solar desalination plant is planned to be installed. Therefore, the global capacity will be 600MW (CSP, PV and hybrid power plants) by 2020 [8].

2.6.3. *Biomass.* A great amount of waste is available in Jordan, where the equivalent annual power production is estimated at 698.1 GW h. (about 5 % of the consumed electricity). Many projects dealing with waste are in operation across the country. Now, concerning, the goal of the production, which the government is planning to produce by 2020 is about 50 MW [9].

2.7. Case of lebanon

Lebanon remains one of the rare countries where the electrification rate is at 100%, but suffers from frequent daily power blackouts. Knowing that there are no identified fossil fuel resources currently available. The development of renewable energies in the country could be an interesting alternative. Despite all the efforts being made to promote Renewable Energies in Lebanon, these forms of energy are struggling to take their place in the energy market, and they account for as little as 2% [10]. However, the government has an objective to attain an integration of 12 % from renewable in the energy mix by the year 2020.

2.7.1. *Wind.* A wind map was updated in 2011, confirmed the existence of an interesting potential. Few installations that exist are of low power (particular instillations). However the government aims to reach a total of 100 MW of installed capacity.

2.7.2. Solar. For solar energy the production potential is very interesting, Lebanon is a country fortunate enough to have a very high amount of sunny hours, even during winter days, with a daily

Average solar insolation of 4.8 kWh/m^2 [11]. But if we can talk about a market that is growing pretty well for the individual solar water heater; other solar technologies including solar thermal collective or photovoltaic can not really take off, despite some projects in this area. Several reasons are behind this fact, the most important in our opinion is the absence of a clear and coherent policy for the energy sector in Lebanon. Furthermore, the government has a plan to develop PV installations up to 10 MW of capacity by 2020.

3. Conclusion

The availability of sustainable resource in the region must exploited by all the countries in the region for the establishment of a clean and productive economy. This will allow, for sure, having investments less costly and reducing the harmful effects of greenhouse gas emissions on the environment. However, the MENA region, renewable energy technologies can be considered as an economic solution for the supply of their rapidly expanding domestic energy markets. It can also be seen as a source of good quality jobs including local hiring, training and transfer of knowledge. Governments also have the choice to promote renewable energy via a different set of measures known as 'fiscal and regulatory incentives', they may help achieve renewable targets in the short- and medium-term.

4. Acknowledgments

This work was supported within the framework of project "REELCOOP – Research Cooperation in Renewable Energy Technologies for Electricity Generation", co-funded by the European Commission (FP7 ENERGY.2013.2.9.1, Grant agreement no: 608466).

5. References

[1] A. Spetan Kh. A. Al-Salaymeh, M. Mahmoud, M. Elkhayat and A. Abu-Jeries, Friedrich Ebert Stiftung, 1st edit., Amman, Jordan (2015)
[2] http://portail.cder.dz/spip.php?article4565, website consulted on November 2016
[3] http://portail.cder.dz/spip.php?article4445, website consulted on January 2017
[4] GIZ. Moroccan-German Energy Partnership. PAREMA. Technical report. (2016)
[5] http://www.ecomena.org/solar-tunisia/, website, consulted on July 2017.
[6] A. Spetan Kh, A. Al-Salaymeh, M. Mahmoud, M. Elkhayat and A. Abu-Jeries, Friedrich Ebert Stiftung, 1st edit., Amman, Jordan (2015).
[7] https://www.ecomena.org/renewable-energy-in-jordan/, website consulted on October 2017.
[8] http://www.cigre.org.jo/arabic/Kiwan%20Jordan%20CIG%202016f.pdf, website consulted on October 2017.
[9] Z. Alamamre, M. Saidan, M. Hararah, K. Rawajf, H.E, Alkhasawneh, M.Al-Shannag, Ren. & Sus. En. Rev. 67, 295 (2017)
[10] ADEME. Technical report. (2015)
[11] A.Mourtada Mediterranean solar (hot) spot– preparation phase – Lebanon fact sheet; (2011)

IOP Conf. Series: Earth and Environmental Science **154** (2018) 012004 doi:10.1088/1755-1315/154/1/012004

Decentralized and cost-effective solar water purification system for remote communities

Hafiz M. Abd-ur-Rehman[1,*], Sehar Shakir[2], Atta-ur-Razaq[1], Hamza Saqib[1], and Saad Tahir[1]

[1]School of Mechanical & Manufacturing Engineering (SMME), National University of Sciences & Technology (NUST), H-12 Campus, Islamabad, Pakistan.
[2]U.S Pakistan Centre for Advanced Studies in Energy (USPCAS-E), National University of Sciences & Technology (NUST), H-12 Campus, Islamabad, Pakistan.
*E-mail: abd-ur-rehman_@hotmail.com

Abstract. In this study, a modified stepped solar still is proposed for water desalination. The overall objective of this work is to develop and test the proposed still design to identify the productivity enhancement as compared to conventional basin type solar still. The proposed design takes the advantage of its stepped configuration that allows the water stream to maintain a minimum desirable water column height and the water flow through the stages under the force of gravity. A minimum water depth in the still results in a higher rate of evaporation. The still is also incorporated with Fresnel lens to increase the water temperature that eventually increases the rate of water evaporation. Another important aspect of this design is the incorporation of phase-change-material (PCM) to increase the operational hours of the solar still. Consequently, daily productivity of fresh water is increased.

1. Introduction

Water is a natural resource that is a necessity of all living organisms. Water sources are normally brackish and contain harmful microbes such as bacteria which make water unsafe and unsuitable for drinking. Desalination using solar energy is one of the methods for purifying water. Desalination is a process of removing various minerals and slats from water. Water desalination is done to convert salty and brackish water to fresh water which can be used for drinking and other purposes such as irrigation. Solar still is the oldest solar distillation techniques for water purification. The major disadvantage of the conventional solar still is its low productivity that can be improved by effective design modifications in the conventional solar still. Numerous research works has been done with the goal to improve the performance, productivity and efficiency of the solar stills. Various parameters that influence the productivity include depth of water, the temperature of water at the inlet to still, free surface area of water, water glass temperature difference, wind velocity and solar radiation. Researchers around the globe have worked intensively on different solar still designs. Several modifications have been implemented and studied thoroughly.

Abu-Hijleh et al. [1] investigated the performance of a solar still with different size sponge cubes placed in the basin. The increase in fresh water output ranged from 18% to 273% compared to identical stills without sponges under identical working conditions. Their study proved that the daily output can be improved by incorporating sponges. Velmurugan et al. [2] incorporated fins in the solar still basin, due to which the production rate accelerated. Exposure area was further increased by utilizing sponges. The results were compared with ordinary basin type still. It was found that 15.3% productivity increased when sponges were used and 45.5% increased when fins were used. Sadhana et al. [3] made an effort to investigate the performance of finned basin solar still. In order to obtain a fair comparison, they fabricated two similar double sloped solar stills. The absorbing plate of one of the solar stills is incorporated with fins in order to enhance the heat transfer rate from the basin liner to water, while another still base is flat. It was proved that the daily output of clean water from the finned basin solar still is 18% more than the conventional double slope solar still. Alaudeen et al. [4] in their work, used a stepped solar still to enhance the productivity of the conventional still. They integrated the stepped configuration with inclined flat plate collector. The setup consists of subsequent trays and flat plate collectors. This modification increases the exposure area which enhances the rate of evaporation. Kumar et al. [5] investigated the performance of a single basin double slope solar still in conjunction with a phase change material (PCM). Paraffin wax was used as the thermal energy storage medium under the liner of the basin. Overall 61% gain was recorded when Phase change material was used, where as a 34% nocturnal and 64% gain in the day time output productivity was recorded, which is a drastic improvement. El-Samadony et al. [6] carried out an experimental study on a modified stepped solar still with internal and external reflectors and an external condenser. Results were obtained with external condenser alone as well as with the reflectors. A detailed comparison between the modified-stepped solar still and the conventional setup was conducted. It was determined that the productivity of the modified still with the external condenser alone was 66% higher than that of the conventional solar still. Pandey et al. [7] conducted an experimental study in order to determine the effect of a separate condenser on the performance of a single slope solar still. They carried out a complete comparative study between a conventional solar still and one with a separate condenser. Their results showed that the basin type solar still with separate condenser was 19% more efficient as compared to the conventional type. Pandey [8] studied the effect of bubbling of ambient air and that of simultaneous air bubbling and cooling the glass cover. Bubbling of ambient helps to increase the rate of evaporation [9-11], thus increasing both the productivity and the efficiency.

The literature review inferred that the solar distillation is an appropriate choice for a decentralized small-scale water desalination systems, especially in remote regions where inexpensive land and abundant solar radiations are available. Therefore, a novel design of a bubbler solar still is developed and tested to identify its productivity enhancement as compared to conventional solar still. One major advantage of this system is its stand-alone/off-grid functionality for electricity scarce remote areas.

2. Experimental setup

The 3D design and schematics of experimental setup is shown in figure 1 and figure 2. It consists of a feed water tank that not only store saline water, but also acts as a secondary solar still. A fraction of water in the storage tank evaporates by solar irradiance. The water vapors condense on the top and side glass surfaces of the tank. The distillate is collected in a measuring cylinder attached to the tank. This tank also incorporates phase change material (paraffin wax) below the basin, which act as a medium to store thermal energy as latent heat to provide higher operating hours for the system. The base of the tank is adequately insulated using glass wool to prevent heat losses. A K-type thermocouple installed within this tank will continuously monitor the temperature of the water. The tank is properly sealed from all sides using silicone gel. All the stated modifications in the storage will ensure maximum productivity of clean water from secondary solar still. The pre-heated water from the storage tank is send to the main still via a stainless steel connecting pipe. This pipe incorporates a manually operated valve to control the flow rate of water as it flows from the storage tank to the main still. All the connection associated with this pipe are properly sealed to prevent any leakages. Main solar still is coupled with Fresnel lens to that concentrate solar radiations on the black absorber trays to maximize the thermal energy input in to the still. The trays are configured in a stepped-sloped configuration. A clear glass cover with a tilt angle equal to Islamabad latitude angle (33.7 degrees) is chosen and it will face south to receives maximum amount of solar irradiance. The stepped-sloped tray configuration is supported on a system of multiple threaded steel rods in the still for load distribution. The tray configuration is painted black to maximize the absorption of solar irradiance. Paraffin wax is incorporated as phase change material (PCM) below the basin to act as thermal heat storage medium. PCM provide effective night time functionality for the setup. A mesh of Aluminum fins is distributed within paraffin wax to increase heat transfer from the still base to the wax when solar irradiance is high and from the wax back to the tray base when solar irradiance is low. The base of the still is properly insulated by using a combination of different insulation materials including saw dust and packing material to prevent heat loss from the base of the still. The trays incorporate sponges that act like baffle plates so that water remains in the tray for a longer duration. The sponges also absorb water and increase surface area for evaporation. The inclined trays incorporate black painted aluminum fins which enhance the surface area for heat transfer and act as barriers to slow down the flow rate of water as it flows through the still. The last tray incorporates a perforated pipe through which air is blown to form bubbles in the hot water coming from the stepped-sloped configuration of the main solar still. The bubbling of air through hot water greatly increases the rate of evaporation. Fraction of water evaporated in the main still condenses on the top and sides of the mail still. The distillate is collected in a graduated beaker attached to the mail still. Majority of the water vapors in the main still is send to the external condenser. The condenser contains a coiled copper pipe through which saline water is passed before sending it to the storage tank. The vapors from the main still condense on copper coil surface and release the latent heat of condensation that is used to preheat the seawater. The distillate from the condenser is collected in a graduated beaker attached at the outlet of the condenser. The preheat water is send to the storage tank to complete the cycle.

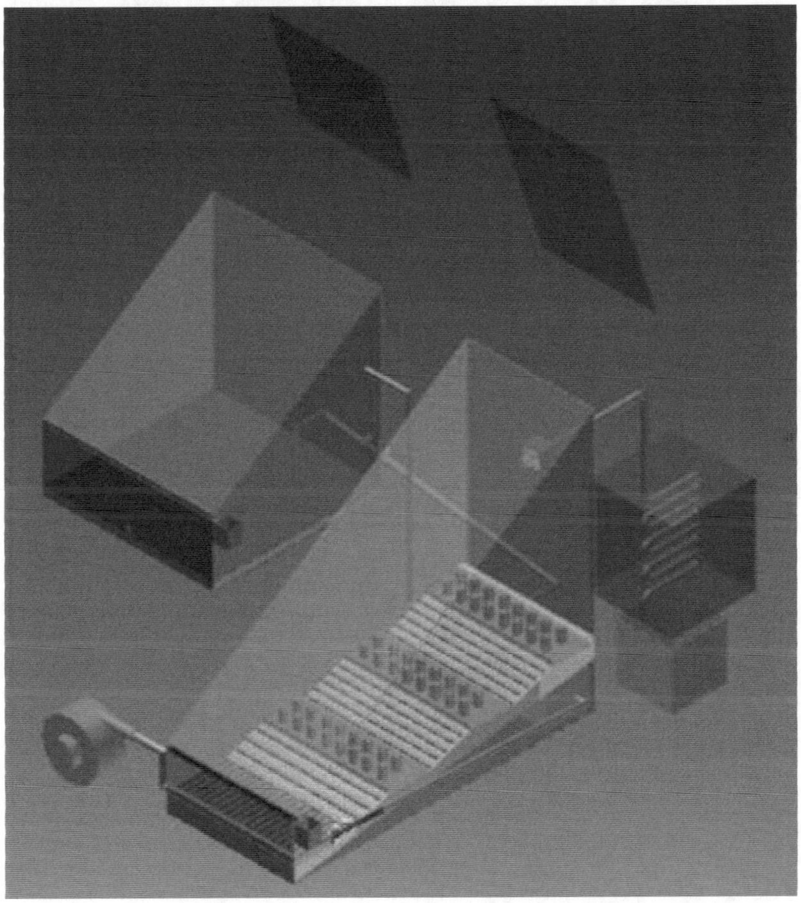

Figure 1: 3D design of the experimental setup.

3. Results and discussion

Experimental analysis was performed along with numerical analysis. The results for both types were computed, plotted and compared. Cumulative productivity of clean output water in L/m^2 was calculated for a 2 cm water depth over a period of 10 hours from 9:00 am to 7:00 pm. The output productivity rate was maximum between 13 to 15 hours and the total productivity obtained from numerical analysis was 6.31L/m2.The resulting graph is as shown in figure 3.

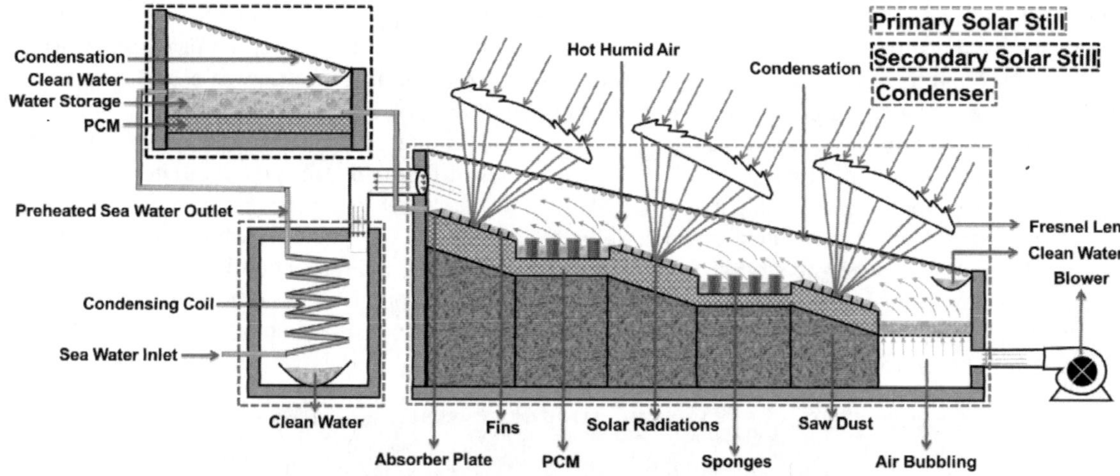

Figure 2: Schematic diagram of the experimental setup of modified solar still.

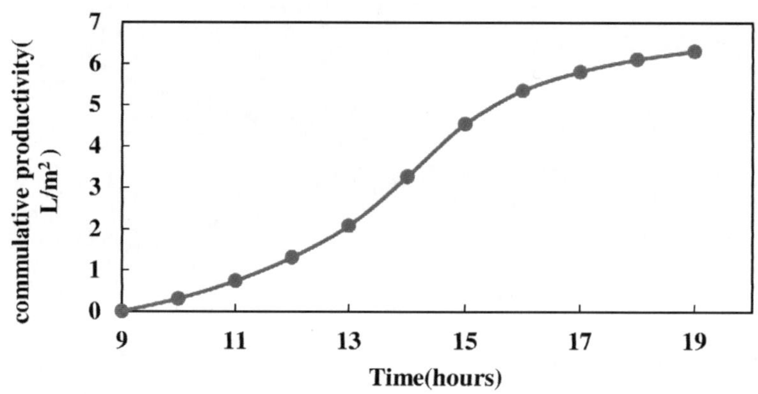

Figure 3: Cumulative productivity in L/m^2 for 2cm water depth

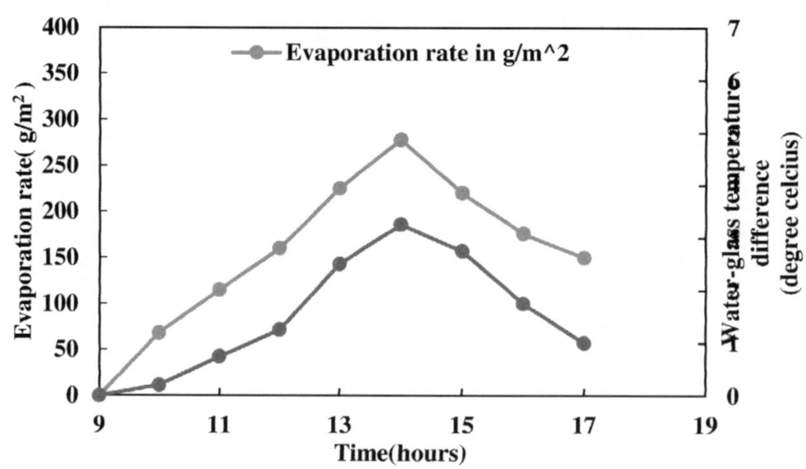

Figure 4: Effect of water-glass temperature difference on productivity.

IOP Conf. Series: Earth and Environmental Science **154** (2018) 012004 doi:10.1088/1755-1315/154/1/012004

The effect of the water-glass temperature difference on the Productivity (Evaporation rate) of output water over a period of 8 hours from 9:00am to 5:00pm was determined through numerical analysis and the results were plotted as shown in figure 4.
The evaporation rate (productivity) in g/m^2 was plotted on the left-hand y-axis while the water-glass temperature in degree Celsius was plotted on the right-hand y-axis along with time in hours on the x-axis. It was found out that the two aspects namely evaporation rate and temperature difference are in direction relation. As the water- glass temperature difference increases, the productivity increases as well. The maximum productivity obtained was 280g/m^2 for a temperature difference of 3.2 degree Celsius at 1400 hours.
Experimental analysis was performed on the final prototype over a time of 10 hours from 9:00am to 7:00pm. The maximum hourly productivity was obtained between 1:00pm and 2:00pm as 0.86 Liters. The hourly productivity of clean output water was noted and the results were plotted in a bar-chart as shown figure 5.

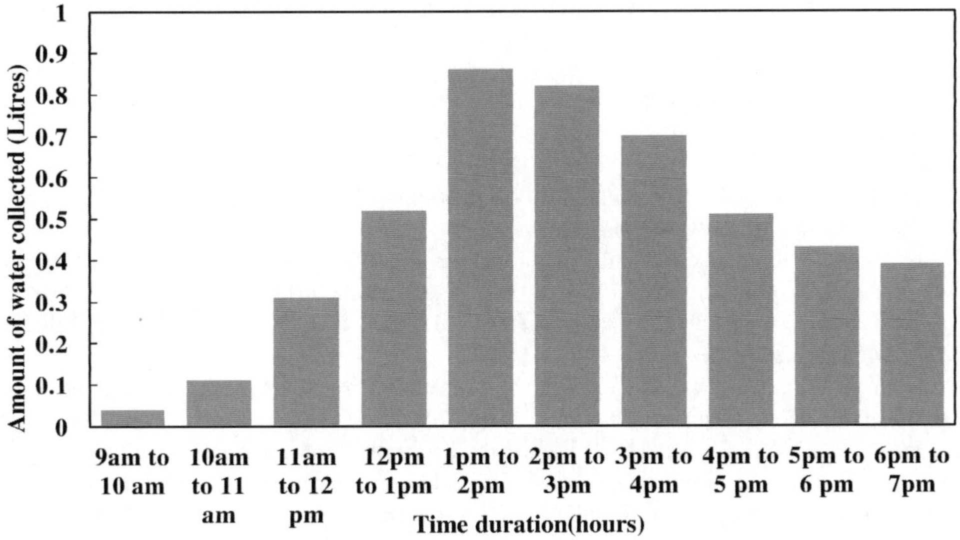

Figure 5: Experimental productivity analysis

Figure 6 shows a comparison between the cumulative productivity determined from numerical analysis and that obtained from the experimental analysis. The productivity obtained from experimental analysis is lower as compared to that obtained from numerical analysis due to heat losses and inefficiencies.

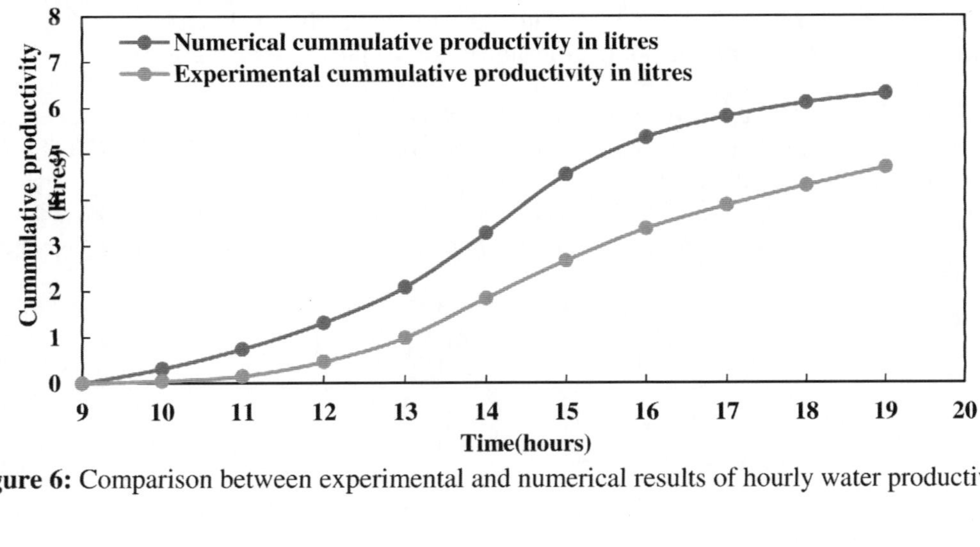

Figure 6: Comparison between experimental and numerical results of hourly water productivity.

4. Conclusion
A modified stepped solar still is designed, developed, and tested in efforts of developing a lost cost decentralized small scale water purification system for remote communities. The stepped configuration in the still are incorporated with sponges to increase the surface are for water evaporation. The pebbles are used in each stage to store the sensible heat and Phase change material (Paraffin Wax) is incorporated to extend the working hours of the solar still. Numerical analysis was conducted using Matlab to determine the effect of various modifications on the productivity of the still. The results show that the maximum productivity was achieved between 12:00-14:00 hours of the day. The cumulative productivity at the end of the day was 6.3 liters.

The developed design is original in its stand-alone/off-grid functionality in electricity scarce remote areas. The results of the proposed work will be a valuable reference for both researchers and engineers in the area of water desalination. It will also alleviate the sustainability concerns of conventional energy based water desalination systems by balancing the potable water requirements through renewable energy source. Moreover, it will provide the prospects of reducing the clean water scarcity in remote areas.

5. Acknowledgements
The authors acknowledge the support of National University of Sciences and Technology (NUST), H-12 Campus, Islamabad, Pakistan.

6. References
[1] H. M. Rababa'h, (2003). Experimental study of a solar still with sponge cubes in basin. Energy conversion and Management, 44(9), 1411-1418.

[2] V. Velmurugan, M. Gopalakrishnan, R. Raghu, & K. Srithar, (2008). Single basin solar still with fin for enhancing productivity. Energy Conversion and Management, 49(10), 2602-2608.

[3] V. Pal, A. K. Rai, & V. Sachan, (2015). PERFORMANCE STUDY OF A SOLAR AIR HEATER. *International Journal*, 6(7).

[4] A. Alaudeen, K. Johnson, P. Ganasundar, A. S. Abuthahir, & K. Srithar, (2014). Study on stepped type basin in a solar still. *Journal of King Saud University-Engineering Sciences,* 26(2), 176-183.

[5] A. Kumar, A. K. Rai, & R. Garg, (2015, February). Experimental investigation of a passive solar still with paraffin wax as latent heat storage. In Technologies for Sustainable Development (ICTSD), 2015 International Conference on (pp. 1-6). IEEE.

[6] Y. A. F. El-Samadony, A. S. Abdullah, & Z. M. Omara, (2015). Experimental study of stepped solar still integrated with reflectors and external condenser. Experimental Heat Transfer, 28(4), 392-404.

[7] Narayan Pandey and Dr. A. K. Rai, Performance Study of Solar Still with Separate Condenser. *International Journal of Mechanical Engineering and Technology,* 7(4), 2016, pp. 125–130.

[8] G. C. Pandey, (1984). Effect of dried and forced air bubbling on the partial pressure of water vapour and the performance of solar still. Solar energy, 33(1), 13-18.

[9] H. M. Abd-ur-Rehman, & F. A. Al-Sulaiman, (2017). A novel design of a multistage stepped bubble column humidifier for the humidification of air. Applied Thermal Engineering, 120, 530-536.

[10] F. A. Al-Sulaiman, H. M. Abd-ur-Rehman, M. A. Antar, & O. Munteshari, (2017). Experimental investigation of a bubble column humidifier heated through solar energy. DESALINATION AND WATER TREATMENT, 60, 58-69.

[11] H. M. Abd-ur-Rehman, & F. A. Al-Sulaiman, (2016). An experimental investigation of a novel design air humidifier using direct solar thermal heating. Energy Conversion and Management, 127, 667-678.

IOP Conf. Series: Earth and Environmental Science **154** (2018) 012005 doi:10.1088/1755-1315/154/1/012005

Performance degradation of photovoltaic modules at different sites

A Hadj Arab[1,a], I Hadj Mahammed[2], S Ould Amrouche[1], B. Taghezouit[1] and N. Yassaa[1]

[1]Centre de Développement des Energies Renouvelables, CDER, 16340, Algiers, Algeria
[2] Unité de Recherche Appliquée en Energies Renouvelables, URAER, Centre de Développement des Energies Renouvelables, CDER, 47133, Ghardaïa, Algeria
[a] E-mail : a.hadjarab@cder.dz

Abstract. In this work are presented results of electrical performance measurements of 120 crystalline silicon PV modules following long-term outdoor measurements. A set of 90 PV modules represent the first grid-connected photovoltaic (PV) system in Algeria, installed at the level of the "Centre de Développement des Energies Renouvelables" (CDER) site (Mediterranean coast), Bouzareah. The other 30 PV modules were undertaken in an arid area of the desert region of Ghardaïa site, about 600 km south of Algiers, with measurements collected from different applications. Following different characterization tests, we noticed that the all tested PV modules kept their power-generating rate except a slight reduction. Therefore, a mathematical model has been used to carry out PV module testing at different irradiance and temperature levels. Hence, different PV module parameters have been calculated from the recorded values of the open-circuit voltage, the short-circuit current, the voltage and current at maximum power point. The electrical measurements have indicated different degradations of current-voltage parameters. All the PV modules stated a decrease in the nominal power, which is variable from one module to another.

1. Introduction

The decrease in the cost of crystalline silicon photovoltaic (PV) modules is noticeable and embraces a great promising future. However, it raises the question of whether weather conditions can these PV modules sustain.

The manufacturers deploy efforts in order to increase the PV modules lifetime, even under harsh operating conditions. Therefore, the accelerated and qualification tests procedures have been used to anticipate the aging mechanisms, and to establish a realistic standard quality. When a PV-module is bought, it is often accompanied with a warranty of more than 20 years, which is based on a type certification IEC 61215 and proceeded with a number of accelerated lifetime tests. Therefore, it is recommended to support the accelerated tests data with data from field-testing, in order to prove the users that a warranty of more than 20 years is valid. Substantial progress has been achieved, but more research is needed and then manufacturers will have to establish serious standard quality control in order to fulfill the long-term goals. Hence, to predict lifetime, mainly in remote areas, it is necessary to study the aging of the PV modules in various climatic areas.

Jordan and Kurtz examined degradation rates of flat-plate terrestrial modules and systems, from field-testing throughout the last 40 years. Approximately 2000 degradation rates, measured on individual modules or entire systems have been collected from and show a mean degradation rate of 0.8%/year

and an average value of 0.5%/year. The majority, 78% of all data, stated a degradation rate of <1%/year. This cumulative field experience may now support long-term warranties [1]. Kichou et al. [2] have studied the degradation of the micro-morph Thin Film Photovoltaic. (TFPV) modules in a relatively dry and sunny inland site with a Continental-Mediterranean climate in Jaén, Spain. They have noted the reduction of the DC output power of about 12.51% after the first month of exposure under outdoor conditions. Furthermore, the stabilization period was observed to start after four months of operation with a total reduction of the PV array DC output power of 16.66%.
Kahoul et al. [3] have studied the electrical performance degradation of local manufacturing PV modules (UDTS-50), in Algeria. Results reveal the degradation of some PV modules up to 12% compared to their original state. Jordan et al. found that median degradation for x-Si technologies in the 0.5–0.6%/year range with the average range of 0.8–0.9% /year. Hetero-interface technology (HIT) and microcrystalline silicon (μc-Si) technologies even though are not as abundant, exhibit degradation around 1%/year and resemble thin-film products more closely than x-Si. On the other hand, they mentioned that nonlinearities for some products have been documented in some studies for the worst performing modules and are influenced by a number of factors. Modeling these nonlinearities, whether they take place at the beginning-of-life or end-of-life in the PV life cycle, has a significant effect on the LCOE [4].
A first study dealing with the measurements of electrical performance of crystalline silicon PV modules following long-term outdoor continue exposure was carried out by Hadj Arab et al. [5]. They noticed that all the modules presented a decrease of the nominal power.
In our case, we aim at evaluating the degradation of the electrical characteristics of 120 PV modules, tested at different sites. 90 PV modules of monocrystalline silicon technology operating, at CDER, Bouzareah with a Mediterranean climate condition, and 30 modules of monocrystalline at Ghardaïa site of arid climate. The period of exposure of these modules are from June 2004 for the PV system of Bouzareah site and from July 2008 for the site of Ghardaïa.

2. Photovoltaic array

The total number of the PV modules previously mentioned forming the two PV generators (PVG1 and PVG2) and operating for different applications have been tested according to the type of use, as follows:

2.1. Photovoltaic array PVG1

The PVG1 array consists of 90 PV modules; it was installed on a horizontal roof at the CDER building (Algiers). This system was acquired within the CIEMAT cooperation (Spain). The station's latitude is 36.8°. The unique tilt angle considered of the PV array is 25°, which favors energy production in the summer time. Note that Algiers is near the sea and is affected by the seasons [6]. Due to the excessive use of air conditioners, the peak loads occur on summer afternoons in Algeria.

Table 1. The technical characteristics of the used PV module in PVG1

Parameter	Value
Width (mm)	654
Length (mm)	1310
Weight (kg)	11.5
Number of cells in series	36
Number of cells in parallel	2
Maximum Power (Wp)	106
Short-circuit current (A)	6.54
Current at Maximum Power Point (A)	6.10
Open circuit voltage (V)	21.6
Voltage at Maximum Power Point (V)	17.4

IOP Conf. Series: Earth and Environmental Science **154** (2018) 012005 doi:10.1088/1755-1315/154/1/012005

The PV array consists of 90 PV 106 Wp crystalline silicon modules (see table 1). It is composed of three sub-arrays of 30 modules each. The sub-array configuration is 15Sx2P. Each sub-array is connected to a 2.5 kVA inverter [7].

2.2. Photovoltaic array PVG2

Similar to the PVG1, the current PVG2 array, which consists of 30 PV modules of the same technology, was installed on a roof at URAER (Applied Research Unit for Renewable Energy) building (Ghardaïa), located at a latitude of 32.4 °. The common tilt angle was close to the latitude angle of Ghardaïa. Huge irradiances and temperatures characterize this site.

The PV array consists of the two following sub systems:

- Subsytem1 consists of 10 PV modules of 100 Wp, connected in (1Sx10P) configuration (see table 2).
- The subsystem 2 consists of 20 PV modules of 50 Wp configured in (2Sx10P).

Table 2. The technical characteristics of the used PV module In PVG2

Parameter	First sub PVG2	Second sub PVG2
Width (mm)	654	340
Length (mm)	1310	1304
Weight (kg)	11.5	5.5
Number of cells in series	36	36
Number of cells in parallel	2	1
Maximum Power (Wp)	100	50
Short-circuit current (A)	3.27	3.27
Current at Maximum Power Point (A)	2.87	2.87
Open circuit voltage (V)	43.2	21.6
Voltage at Maximum Power Point (V)	34.8	17.4

All the subsystems are connected to a 2 kVA inverter and supply a stand-alone PV system. Table 2 gives technical characteristics of these two subsystems respectively.

3. Modeling of PV module

A PV array is a nonlinear power source. At a given irradiance, the current-voltage (I-V) relationship is illustrated by:

$$I = I_{ph} - I_s\left[exp\left(\frac{V + R_s I}{mV_t}\right) - 1\right] - \frac{V + R_s I}{R_{sh}} \quad (1)$$

Where $V_t = \frac{k.T}{e}$

I_{ph} is the photocurrent, I_s the saturation current, Rs the series resistance, m the diode quality factor, R_{sh} the shunt resistance, V_t the thermal voltage, k the Boltzmann's constant, T the temperature of the solar cell (K) and e the charge of the electron.

The analytical five-parameter method has been used to measure the PV array performances at different irradiance and temperature conditions [8].

At a particular temperature and illumination, the five parameters I_{ph}, I_s, R_s, R_{sh} and m can be computed from the open-circuit voltage (V_{OC}), the short-circuit current (I_{SC}), the voltage and current at maximum power point (V_m, I_m), R_{s0} and R_{sh0}.

$$\left(\frac{dV}{dI}\right)_{V=V_{0c}} = -R_{s0} \quad (2)$$

$$\left(\frac{dI}{dV}\right)_{I=I_{sc}} = -R_{sh0} \quad (3)$$

Where: R_{s0} and R_{sh0} are reciprocal of slope at open circuit point and reciprocal of slope at short circuit point respectively.
The obtained equations are:

$$m = \frac{V_m + I_m R_{s0} - V_{OC}}{V_t\left[\ln\left(I_{SC} - \frac{V_m}{R_{sh}} - I_m\right) - \ln\left(I_{SC} - \frac{V_{OC}}{R_{sh}}\right) + \left(\frac{I_m}{I_{SC} - \frac{V_{OC}}{R_{sh0}}}\right)\right]} \tag{4}$$

$$I_s = \left(I_{SC} - \frac{V_{OC}}{R_{sh}}\right)\exp\left(-\frac{V_{OC}}{mV_t}\right) \tag{5}$$

$$R_s = R_{s0} - \frac{mV_t}{I_s}\exp\left(-\frac{V_{OC}}{mV_t}\right) \tag{6}$$

$$I_{ph} = I_{SC}\left(1 + \frac{R_s}{R_{sh}}\right) + I_s\left(\exp\frac{I_{SC}R_s}{mV_t} - 1\right) \tag{7}$$

$$R_{sh} = R_{sh0} \tag{8}$$

Equations (4)-(8) have been used to estimate the five parameters I_{ph}, I_s, R_s, R_{sh} and m that are necessary to apply equation (1).
In order to show how accurately the model works, measured current-voltage curves for a (Si) module are compared with modelled curves. The measured and simulated curves have been found to agree well [5].

The curve of equation 1 is only applicable at one particular irradiance (G_1) and temperature (T_1). A model is used to translate this curve to other irradiance (G_2) and temperature (T_2) according to (9)-(14): [8]

$$I_{SC2}(G_2,T_2) = I_{SC1}(G_1,T_1).\frac{G_2}{G_1} + \alpha(T_2 - T_1) \tag{9}$$

$$V_{OC2}(G_2,T_2) = V_{OC1}(G_1,T_1) + mV_t \cdot \ln\left(\frac{G_2}{G_1}\right) + \beta(T_2 - T_1). \tag{10}$$

$$I_2 = I_1 + \Delta I_{SC} \tag{11}$$

$$V_2 = V_1 + \Delta V_{OC} \tag{12}$$

$$\Delta I_{SC} = I_{SC2} - I_{SC1} \tag{13}$$

$$\Delta V_{OC} = V_{OC2} - V_{OC1} \tag{14}$$

Where α is temperature coefficient of current (A/K) and β is temperature coefficients of voltage (V/K). In order to evaluate this model, we have compared measured I-V curves at a variety of ambient conditions, with the curves defined by equations 1 to 14. A good fit between the measurement and simulation is obtained.

4. Degradation rate

In order to determine the degradation rate (DR) of the maximum power output (*P*max) of the PV modules, measured standardized values are matched to data of the PV module given by the manufacturer. The difference in percentage represents the reduction of the parameter. The degradation rate of parameters is given respectively by the following equation [9]:

$$DR(\%) = \left(\frac{P\max_0 - P\max}{P\max_0}\right) \times 100 \tag{15}$$

Where: $P\max_0$ and $P\max$ are respectively the maximum power given by the manufacturer and the maximum measured power

5. Results and discussions

5.1. Photovoltaic array PVG2

The PVPM2540C device (figure 1.a) provide the I-V curve measurement with 101 samples in 2 seconds for photovoltaic modules and small strings up to 250V and 40Adc , the device measure the irradiance and PV cell temperature using sensors box (figure 1.b).

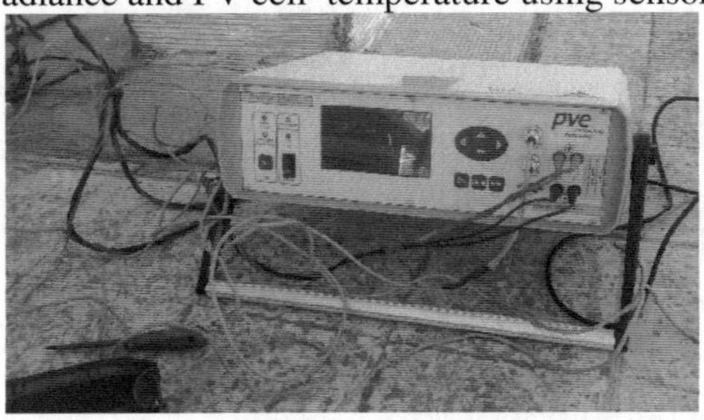

Figure 1.a. I-V curve measurement device PVPM2540

Figure 1.b. Irradiance & temperature sensors box fixed on the plane of the photovoltaic modules tested

Each PV module was washed to eliminate the dirt and the sand gathered on the front glass due to long-term field exposure.

Figures 2 illustrates the degradation of each PV module of the three sub-arrays. These degradations are divided into three sets:

moderate degradation: 0.5- 5% /year □

average degradation: 5-10% /year ▒

high degradation: ≥10% /year ■

Figure 2.a shows the degradation of sub-array G1 with 3.92 % /year of DR. It can be seen that the majority have moderate degradation. Only 30% of this sub-array present an average degradation, particularly in the second string (S2). Unlike G1, the sub-array G2 holds the poorest performance between the three sub arrays (Figure 2.b), with 8.52% of DG /year. G2 contains the worst DR 13.75% /year for a module situated in string 2 due to the clear crack of one cell in the PV module.

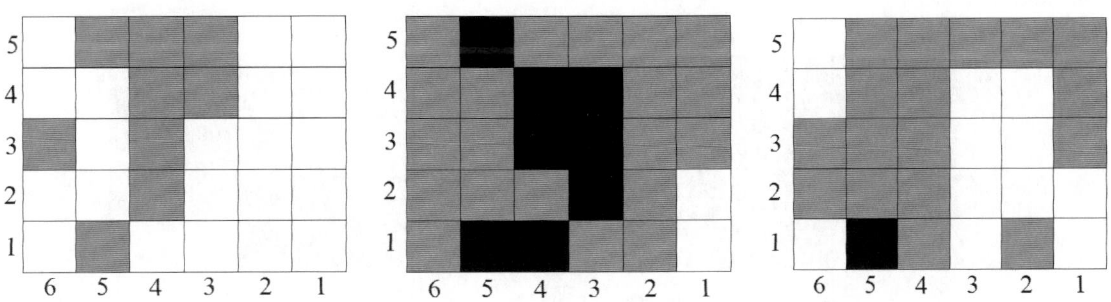

Figure 2. Degradation rates of PVG1 sub-arrays : a) Sub1, b) Sub2 c) Sub3

The next figure 3 presents a detailed estimation of modules degradation for the three sub arrays.

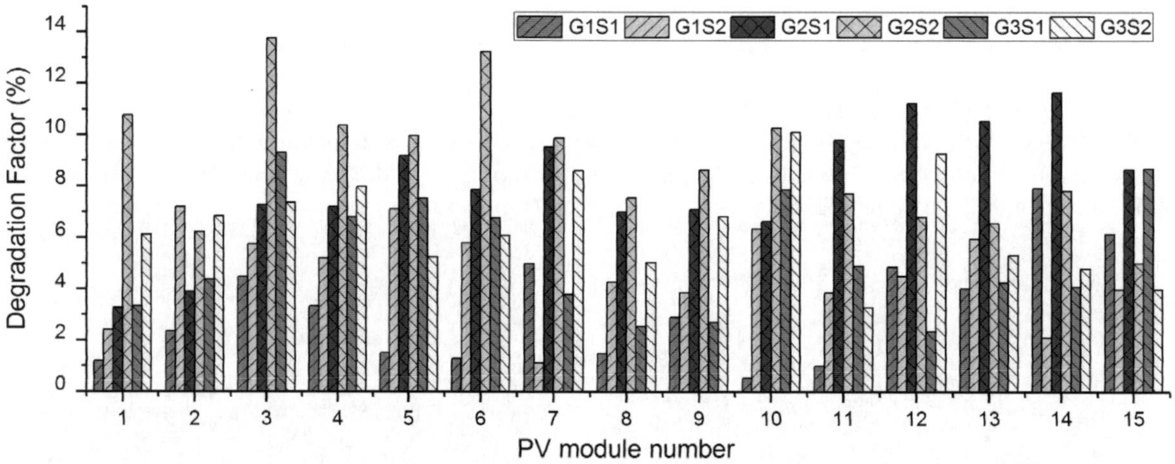

Figure 3. Degradation factor of PVG1 modules

Table 3 summarizes the minimum, average and maximum performance of all modules of the installed grid-connected PV system. It provides the maximum power modules for different performances reported at STC conditions: their degradation rates and their positions in the PV generators. Thus, the best performance is assigned to a module from the sub array1; also, the medium to one module of the third sub array3 and the worst performance belongs to the second sub array2.

Table 3. Performance of all modules of the installed grid connected PV system PVG1

Parameters	Minimal Performance	Average performance	Maximal performance
Pmax (W)	91.43	99.54	105.44
DR (%)	13.75	6.10	0.53
Position	G2; S2; M3	G3; S2; M6	G1; S1; M10

5.2. Results for PVG 2

Figure 4.a presents the degradation of the sub array1 of GPV2 with 7.52 % /year of DR. It can be seen that the majority have moderate degradation, except for one module of this array, which has high degradation with 16.25 % /year of DR. However, 90% of this array has average degradation between 5.07% and 8.32 % of DR /year. (See table 4).

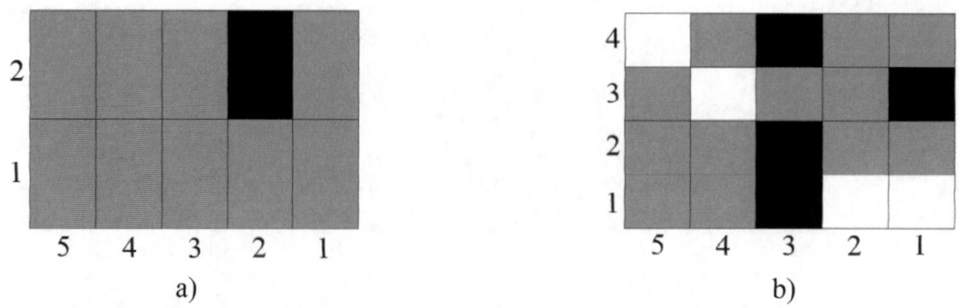

Figure 4. The degradation rates of PVG2 sub-arrays : **a) Sub1, b) Sub2**

The degradation of the sub-array 2 of GPV2 is presented in Figure 4.b. 7.19 % /year of DR is observed. This degradation can be divided into three categories as follows: 20% of moderate, 20% of medium and 60% of high degradation.

Table 4. The degradation factors of the tested PV modules of the sub- array1 of PVG2.

Sub-array1 (PVG2)		**Sub-array2 (PVG2)**			
	M 1		M 1	M2	String Mean DR
S1	5.65	S1	5.96	10.67	8.32
S2	8.32	S2	5.11	4.72	4.92
S3	16.25	S3	7.91	7.31	7.61
S4	7.03	S4	5.01	1.93	3.47
S5	6.74	S5	21.88	6.5	14.19
S6	7.39	S6	12.95	12.67	12.81
S7	7.37	S7	5.86	2.04	3.95
S8	5.79	S8	5.28	5.73	5.51
S9	5.62	S9	1.08	8.79	4.94
S10	5.07	S10	6.86	5.58	6.22
Mean DR	7.52			7.19	

5.3. Visual inspection

Visual inspection is a part of the test described in IEC61215. It is the first step to evaluate the degradation mode in PV module. In order to present the long-term degradation of PV modules under test, the inspection allows detection some failures after 11 years of exposure in the first site and seven years in the second site that can be observed visually, such discoloration over partial and whole cells, corrosion of string interconnection, loose and brittle of junction boxes.

6. Conclusion

The degradation of the PV modules of the different PV systems installed in two different regions was presented in this work. Outdoor testing was performed on 120 modules, 90 of them were installed at Bouzareah (CDER) site on June 2004, and 30 other modules were installed at the site of Ghardaïa on July 2008. The outdoor measurement was carried out in two different climates. The results were translated and compared with the characteristics given by the manufacturers of the tested modules. The modules have an average peak of power degradation between 0.53% and 21.88% lower than the average value measured before exposure.

We can notice that the modules are not only exposed but still working since their installation. However, we observed that the degradation is not homogeneous; the modules located in the middle of the 90 modules of the grid-connected system were the most affected by degradation is probably due to the lack of natural ventilation despite the fact that the modules are exposed to the sea. The higher degradation was observed on the system installed in a semi-arid climate, which is characterized by a hostile climate.

7. Acknowledgment

We would like to acknowledge the EU through the 7th F.P for the financial support of the work under the REELCOOP project (G.A. 608466). Special thanks are due to D. C. Jordan from National Renewable Energy Laboratory (NREL, USA) for his valuable collaboration.

8. References

[1] D. C. Jordan and S. R. Kurtz, « Photovoltaic Degradation Rates — an Analytical Review », Prog. Photovolt. Res. Appl. (2011).

[2] S. Kichou, E. Abaslioglu, S. Silvestre, G. Nofuentes, M. Torres-Ramírez, and A. Chouder, « Study of degradation and evaluation of model parameters of micromorph silicon photovoltaic modules under outdoor long term exposure in Jaén, Spain », Energy Convers. Manag. vol. 120, pp. 109 119 (2016).

[3] N. Kahoul, M. Houabes, and M. Sadok, « Assessing the early degradation of photovoltaic modules performance in the Saharan region », Energy Convers. Manag. vol. 82, pp. 320 326 (2014).

[4] D. C. Jordan, S. R. Kurtz, K. Vansant, and J. Newmiller, « Compendium of photovoltaic degradation rates », Prog. Photovolt. Res. Appl. (2016).

[5] A. Hadj Arab, Y. Bakelli, S. Semaoui, F. Bandou and I. Hadj Mahammed, "Long term performance of crystalline silicon photovoltaic modules," in 3rd International Renewable and Sustainable Energy Conference (IRSEC), Marrakech, pp. 1-5 (2015).
[6] A. Hadj Arab, B. Ait Driss, R. Amimeur, and E. Lorenzo, « Photovoltaic systems sizing for Algeria », Solar Energy, vol. 54, p. 99 104 (1995).
[7] A. Hadj Arab, F. Cherfa, A. Chouder, and F. Chenlo, « Grid-Connected Photovoltaic System at CDER-Algeria », in 20th European Photovoltaic Solar Energy Conference and Exhibition. Barcelona, pp. 6–10 (2005).
[8] A. Hadj Arab, F. Chenlo, and M. Benghanem, « Loss-of-load probability of photovoltaic water pumping systems », Solar Energy, vol. 76, N° 6, pp. 713 723 (2004).
[9] F. Bandou, A. Hadj Arab, and M. S. Belkaid, « Test Performance Electrical of the Photovoltaic Module in Two Different Environments », Energy Procedia, vol. 36, p. 1018 1024 (2013).

IOP Conf. Series: Earth and Environmental Science **154** (2018) 012006 doi:10.1088/1755-1315/154/1/012006

The impact of predicted demand on energy production

I. El kafazi[1], R. Bannari[2], My. O. Aboutafail[3]

[1,2]Laboratory Systems Engineering, Ensa, Ibn Tofail University Kenitra, Morocco
[3]Laboratory of Electrical Engineering Computing Sciences, Ensa, Ibn Tofail University Kenitra, Morocco

Abstract. Energy is crucial for human life, a secure and accessible supply of power is essential for the sustainability of societies. Economic development and demographic progression increase energy demand, prompting countries to conduct research and studies on energy demand and production. Although, increasing in energy demand in the future requires a correct determination of the amount of energy supplied. Our article studies the impact of demand on energy production to find the relationship between the two latter and managing properly the production between the different energy sources. Historical data of demand and energy production since 2000 are used. The data are processed by the regression model to study the impact of demand on production. The obtained results indicate that demand has a positive and significant impact on production (high impact). Production is also increasing but at a slower pace. In this work, Morocco is considered as a case study.

1. Introduction

The energy demand should increase considerably in the coming years because of the increase of the population and the economic development (EIA, on 2007). Gradually as economics spend some concept of subsistence to that of industrial production or services, the lifestyles of numerous individuals will know of deep changes. The main increases of energy demand will be the fact of the developing countries where the percentage of the world consumption of energy should cross 46 to 58 percent between 2004 and 2030 (EIA, on 2007). It is planned that the energy consumption in developing countries will increase at an annual average rate of 3 percent between 2004 and 2020. In industrialized countries, where the state economies are mature and the population growth should be relatively weak, the energy demand will progress more slowly at the 0, 9 percent rate a year, although the starting point is much higher. The increase of the world demand for 2030 will be attributable in the electricity production and the fifth at the needs for transport, in particular in the form of petroleum-based fuels (EIA on 2007). The electrical energy can be an important element of any strategy wished to improve the environment provided that the consequences at the level of his production are mastered.

2. Literature review

There are various types of conventional and no conventional sources of energy used for electrical power production. The system of solar and wind energy is one of the most important sources of energy. The exploitation of solar and wind energy became more and more necessary because of the nature modular and favorable to the environment [1]. The modeling of energy demand can follow either the approach with a variation of time or the approach with fixed coefficients. Nevertheless, he can be necessary to estimate the underlying statistical behavior of the parameters of energy demand before any significant inference. Z. Mohamed et al. used the model of multiple linear regression based on the gross domestic product, the price of electricity and the population to forecast consumption in New Zealand [2]. M. Tunç et al. studied and compared the world production of electric energy from different power sources with electricity generation in Turkey [3]. S. Jebaraja et al. presented various types of models used in the literature such as energy planning models, model supply, demand of energy, prediction models, and renewable energy patterns, emission reduction and optimization models [4]. Y. Lee et al. proposed a model that combines Grey residual modification with genetic programming. They concluded that the model is offering a predictive accuracy than other models [5]. On the other hand, a method of energy consumption based on gross domestic product and population by a model of linear regression exhibited by G. Aydin. 2014. The author concludes that several approaches validate the model, and it can be used to predict the energy consumption of Turkey [6]. C. Adjamagbo et al. suggested four mathematical models for modeling the energy demand in Togo [7]. D. Neto Studied diesel demands in Switzerland by an iterative methodology for the long term. The model based on Chebyshev polynomials in two steps for short-term dynamics [8]. Huneke et al. [9] used the linear programming to obtain an optimal configuration for a combination of generator with wind power - solar energy - battery- diesel for two real powers except for network systems in India and Colombia. The results of optimization for both studies show the possible combination of cells- PV-and diesel generator. In this paper the main contributions are: (i) studying the impact of predicted demand on production; (ii) using the linear regression method to find the best results; (iii) the presentation of preliminary simulation results showing that demand has a positive and significant impact on production. This work is the result of the two previous studies [12], [13].

3. Interest of study

As aforementioned, this study is based in Morocco, which has seen an energetic evolution. The main interests of studying the impact of predicted demand on energy production is that to be able of managing and coordinating DGs, storages, and loads in a more decentralized way reducing the need for the centralized coordination and management. Hence, the optimization of production cost is extremely important in order to cost-efficiently manage its energy resources [14]. The population and economic trends followed by a fast increase of the energy consumption (i.e., increase in the electricity intensity) impellent the countries of the world to lead studies on intersection points between the demand and production. This pattern put them in an unusually favorable situation to implement diverse formulate of reduction of the impact, reducing so at least the uncertainties as for their costs and their advantages. Our paper leads a study to determine the impact of the demand on the energy production to improve the energy efficiency. In addition to achieve the following objectives:

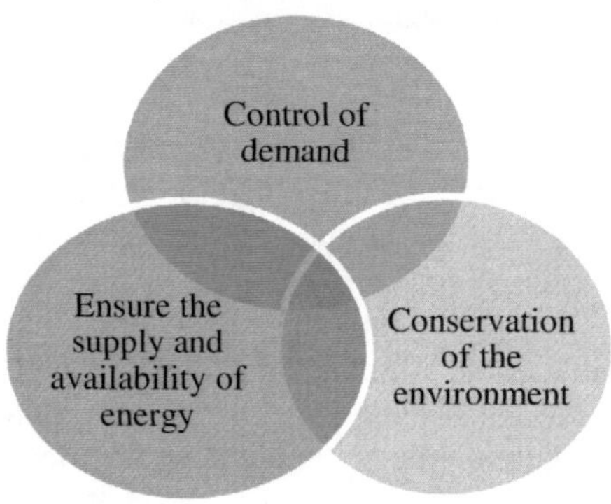

Figure 1. The interest of the study.

4. Methodology

A flow chart for the study is presented (see Fig. 2). This study collects monthly data on Morocco's energy demand and production between 2000 – 2015. Modeling by linear regression is used to examine the impact of demand on energy production in per month in future. In this model, the production is defined as the dependent variable and the demand as an independent variable. Adjustment measures are calculated through R-square and R-square adjusted, and the equation of production determined. The SPSS 20 statistical software which offers a choice of linear regression is used for the analysis. According to the previous studies of demand in [12] and energy production in [13], this represents a first attempt to specifically study the impact of predicted demand on energy production in Morocco. The first target of the present paper is the collecting data, analysis and selecting the methodology. The second target is modeling of the dependent variable as a function of independent variables. The third target is to provide an accurate model for linking demand to production using historical data for demand and energy production.

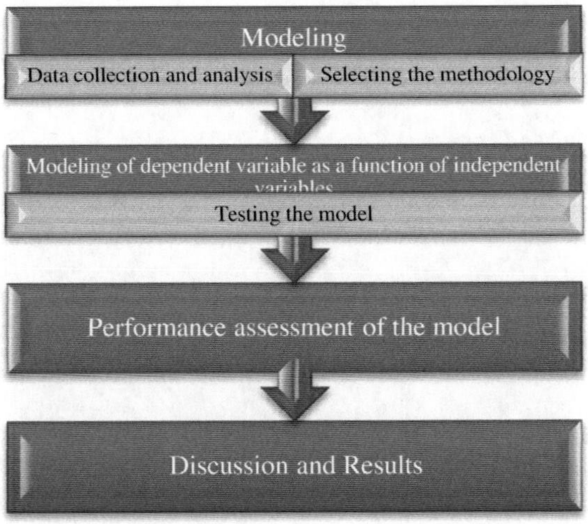

Figure 2. The stages of the study.

The figures below show the evolution of energy demand and production monthly during the 15 years.

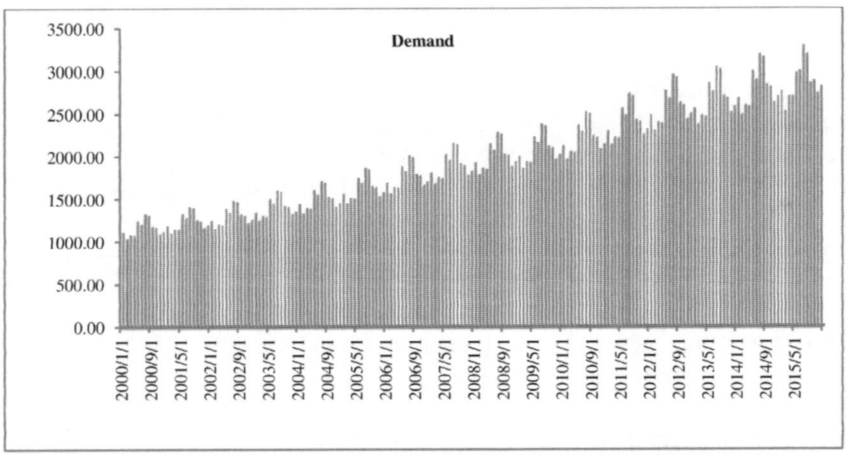

Figure 3. Evolution of energy demand (2000-2015).

From Fig.3 demand decreases seasonally with the temperature when it is cold, whereas higher temperatures increase the electricity consumption due to running cooling devices during the warm periods.

The forecast of energy demand had already studied by the ARIMA model in [12] to determine the needs of consumers in the future years. Therefore, the prediction equation can be formulated as follows:

$$Demand=3.830e^{-6T}-49415.523\pm R \quad (1)$$

With:

D: Demand for energy consumption

T: The time of development in months

R: Includes all elements affecting demand and which are not considered by the model, in addition to the errors in calculating the coefficients.
In this model, the demand is defined as dependent variable, and the time as an independent variable. The model is validated by various statistical approaches such as the determination coefficient, R-square, R-square stationary.
It is concluded that the model can be successfully used as a prediction tool for Moroccan's energy demand [12].

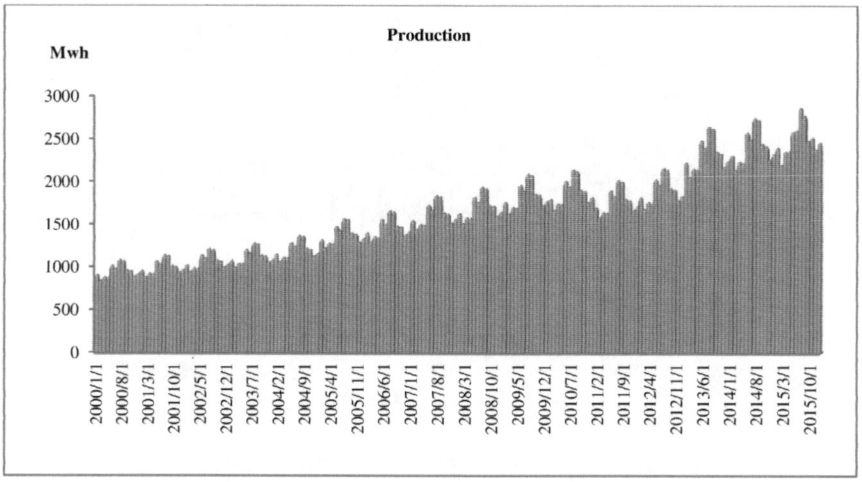

Figure 4. Evolution of energy production (2000-2015).

The forecast of energy production had already studied in [13]. In this case two models are used to predict energy production: polynomial curve fitting and linear regression. The obtained result show that the polynomial curve fitting can be used to predict energy production. In 2012, the power production increased through the construction of new farm of electricity production. Morocco has a considerable potential for electricity generation from the solar energy.
In line with these challenges, this paper focuses on the Moroccan energy and on the potential impact of demand on production. Specifically, this paper explores the effect of demand variable on energy production by analyzing the direct impact of consumption on the electricity production. Hence, demand is a major determinant of electricity production. Moreover, this question is currently of crucial concern given the observed of integrating renewables.
A limitation of our work is that we do not distinguish between residential, commercial and industrial demand for electricity. We quantify the impact of demand variation on energy production only at a national level.

5. Modeling

In this study, different methods were used before choosing the linear regression that gave us satisfactory results.

5.1. Linear Regression

The Regression models are used to study the influence of economic variables on the annual electricity consumption in N. Cyprus is examined [15]. In [16] have examined the nonlinear relationship between energy demand and temperature in the European Union. In [17] examined the dynamic relationship between energy consumption, weather, and price and consumer income. In [18] analyze the linear and nonlinear effect of energy demand on economic growth for Taiwan. The obtained result shows that a threshold regression provides a better empirical model than the standard linear model. The aim of this study is to analyze the difference between the energy demand and production using linear regression for having the anticipation in the future. The integration of renewable energies (e.g., solar and wind) requires a study on the needs of consumers for defining the energy to be produced. Currently,

conventional energy will be unable to cope with future energy requirements. Conventional production (e.g., Fossil fuel and nuclear) are closely linked to environmental deterioration. It is therefore clear that if the increasing energy needs are to be met with preserving the environment.
Renewable energy sources that use natural resources have the potential to provide energy services with almost nil emissions of both air pollutants and greenhouse gases. Renewable energy sources (i.e., solar energy, wind power, biomass and geothermal) are independent, inexhaustible and widely available. These resources can meet the present and future energy demands of the world as presented (see Fig.5).

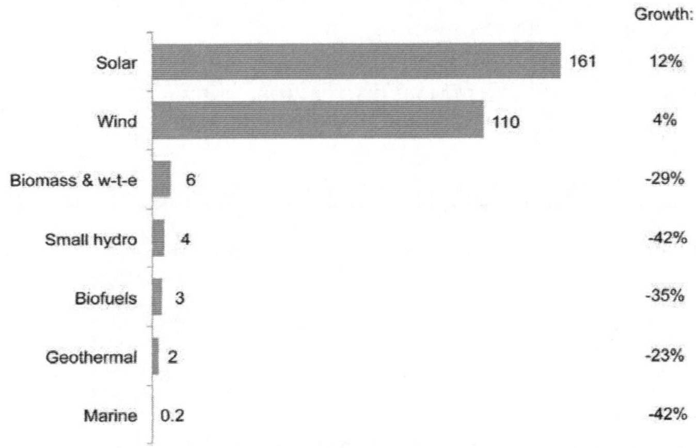

Figure 5. Global investment in renewable energy by technology, developed and developing countries, 2015 [18].

The evolution of renewable energy sources can enhance diversity in energy supply markets, contribute to securing long-term sustainable energy supplies, help reduce environmental impacts and provide commercially attractive options to meet specific energy service needs.
The simple linear regression model is represented by the following equations:

$$y_i = a * x_i + b + \varepsilon_i \tag{2}$$

Where y_i is the value of the dependent variable in observation i, x is independent variable in observation i, a and b are fixed quantities. ε is error term in observation i.

5.2. Test the equality of averages
The probability to know the average of a sample pulled at random can be calculated with the test T-Student. To test this probability for the variables which have in our database. The hypothesis of departure appears as follows:
H0: the average of the studied variables is likely if we pull a sample at random.
H1: the average of the studied variables is not likely if we pull a sample at random.
The results of the test appear as follows:

Table 1. Test on Single Sample

	Test value = 0					
					Confidence interval of difference to 95 %	
Model	t	ddl	Sig. (bilateral)	Average Difference	Inferior	Superior
Demand	46,829	191	0,000	1966,128894187558	1883,3136768	2048,94411147
Production	44,862	191	0,000	1626,418640799437	1554,9095188	1697,92776278

The results of the test T of Student shows that average of the samples is not likely for all the variables of study (test T is very significant for all the variables p - been worth = 0.000) and thus the hypothesis H0 is rejected and H1 is accepted. We conclude that samples possess an extremely improbable average for all the variables of the study.

5.3. Test the normality of variables

To test the normality of variables, the statistics of Kolmogorov-Smirnov for a sample are appealed. The hypothesis of search announce as follows:
H0: the studied variables follow a normal distribution.
H1: the studied variables do not follow a normal distribution.
The results of the test appear as follows:

Table 2. Kolmogorov-Smirnov Test for a Sample

		Demand	Production
N		192	192
Normal settings a,b	Average	1966,128894187558200	1626,418640799437500
	standard eviation	581,7710647158729 00	502,346542018328760
Most extreme differences	Absolute	0,077	0,081
	Positive	0,077	0,081
	Negative	-0,060	-0,062
Test statistics		0,077	0,081
Sig. asymptotic (bilateral)		0,008c	0,004c

a. The distribution of the test is Normal.
b. Calculated from the data.
c. Correction of Lilliefors.

The results of the statistics of K- show themselves that can reject the hypothesis according to which variables follow a distribution according to a law normal (Statistics of the test are very significant for the set of variables p- been worth = 0.000) and the hypothesis which stipulates that the variables of study do not follow a normal distribution can be accepted. (i.e., Rejected H0 and H1 Accepted).

5.4. Information on the regression model

Modeling by Regression is used to analyze the variations of the power production according to the demand. In this model, the variable production is defined as dependent variable, and demand as being an independent variable. Also, the measure of adjustment is calculated through R-square. The study of the impact of the variable "Demand" on the "Production" allowed verifying certain tests to validate the results of regression.
The results of the Test of Homoscedasticity, Test of independence of the terms of the error, the normality of the distribution of the terms of the error and finally the Test of collinearity are satisfactory, and thus we can accept the results of the models.

Table 3. Model Summary

Model	R	R-square	R-square adjusted	Standard Error of Estimate
1	0,979 a	0,959	0,959	101,840413665507740

a. Predicting: (Constant), Demand. b. Dependent variable: Production.

Using the analysis of the table above, gave a good model since "R2 = 0.959", which explains that the explanatory variable contributes almost "95.9%" to the variability of the variable to be explained is production. Thus, the analysis of the results of the F-Fisher test and more particularly the variation of F in order to test whether the most recent contribution presents a significant improvement in the prediction capacity in the regression equation. Important value of the variation of F-Fisher is noticed as it is very significant. This means that the regression equation is good and the explanatory variable contributes very significantly to the scores of the variable production as it is displayed in the table below.

Table 4. Anova[a]

Model		Sum of squares	ddl	Average square	F	Sig.
1	Regression	46228661,948	1	46228661,948	4457,291	0,000b
	Residue	1970579,273	190	10371,470		
	Total	48199241,221	191			

a. Dependent variable : Production
b. Predicting : (Constant), Demand.

The aim of using ANOVA method is for analyzing the influence of the different parameters of linear regression.
The table below shows that there is a statistically impact significant (0.846: p-value = 0.000) of the demand on production. Therefore the demand is the most important determinant for the explanation of production.

Table 5. Model Statistic

Model	Non-standardized coefficients		Standardized coefficients	t	Sig.
	B	Standard Error	Bêta		
(Constant)	-36,224	25,966		-1,395	0,165
Demand	0,846	0,013	0,979	66,763	0,000

The equation linking demand to production can be presented as follows:

$$Production = C + \alpha * Demand \quad + \mu \qquad (3)$$

With:
C: Constant which represents indestructible production
α: Represents the estimated parameter of the relationship between demand and production
μ: Standard Error

The final equation of the regression is:

$$Production = -36.224 + 0.846*Demand \pm 101.84 \quad (4)$$

Demand has a positive and significant impact on production (high impact). When demand increases, the production also increases, but at a slower pace. (i.e., estimated parameter is between 0 and 1).The constant gives a negative sign. The following charts are the graphs that check the hypothesis of Homoscedasticity, the independence of the terms of the error and the normality of the distribution of the terms of the error.

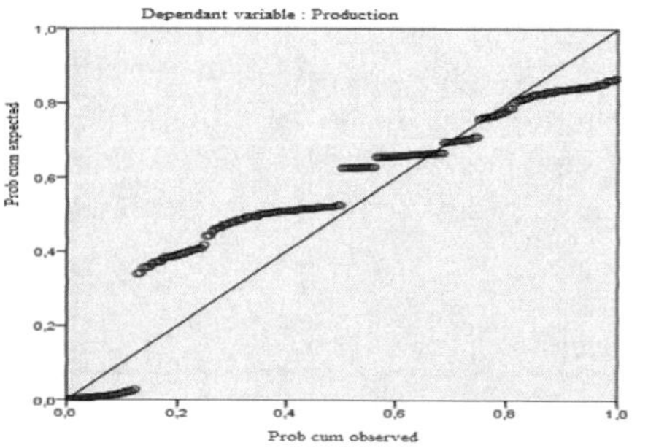

Figure 6. Standardized regression plot

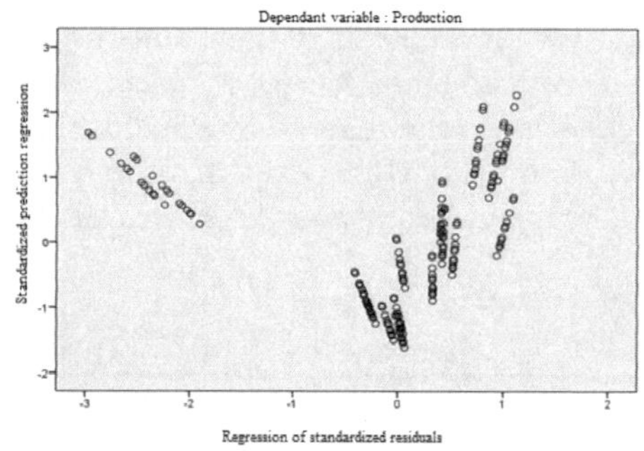

Figure 7. Cloud of point.

In Fig 6 the diagram shows that the distribution of the residues follows the normal distribution, which means that we have a normal distribution of the terms of the error. (i,e. quasi-linear form). In Fig 7 the published scatter of standardized residues in relation to the standardized predicted values does not reveal any particular model, which confirms the assumption of the constant value of the variance of the error term (Homoscedasticity) error.

6. Conclusion

Regression modeling is used to analyze changes in energy production as a function of demand. In this model, the production as a dependent variable and the demand as an independent variable are defined. The fitting measures through R-square and R-square adjusted are calculated. Energy demand is studied with ARIMA model which has a significant result. In this paper, the impact of predicted demand on energy production in Morocco is examined using a linear regression method applied to the last 15 years. Moreover, regression has shown the electricity demand in Morocco, has a positive and significant impact on production. The sensitivity of the demand is higher in the warm period.

In sum, this paper shows that the demand is a major determinant of electricity production in Morocco and that the sensitivity of electricity consumption to temperature has increased in the recent period.

This work allows to mastering the energy transition, satisfies the energy demand and reduction of the energy bill. As well as to assure the management of the energy transition by considering the necessity of protecting the balance of the sector.

7. References

[1] A. Afees, O. Taofeek. Modeling energy demand: Some emerging issues. Renewable and Sustainable Energy Reviews 54 (2016)1470–1480.

[2] M. Tunç, Ü. Çamdali, T. Liman, and A. Değer.Electrical energy consumption and production of Turkey versus world. Energy Policy, VOL. 34, NO 17, 2006. P. 3284-3292, NOV.

[3] S. Jebaraja, S. Iniyanb. A review of energy models. Renewable and Sustainable Energy Reviews. 2006.10 281–311.

[4] Z. Mohamed and P. Bodger, « Forecasting electricity consumption in New Zealand using economic and demographic variables », Energy, vol. 30, no 10, p. 1833-1843, Jul. 2005.

[5] Y.-S. Lee and L.-I. Tong, « Forecasting energy consumption using a grey model improved by incorporating genetic programming », Energy Conversion and Management, vol. 52, no 1, p. 147-152, janv. 2011.

[6] G. Aydin, « Modeling of energy consumption based on economic and demographic factors: The case of Turkey with projections », Renewable and Sustainable Energy Reviews, vol. 35, p. 382-389, Jul. 2014.

[7] C. Adjamagbo, P. Ngae, A. Vianou, and V. Vigneron, « Modélisation de la demande en énergie électrique au Togo », Revue des Energies Renouvelables, vol. 14, no 1, p. 67–83, 2011.

[8] D. Neto. 2012. Testing and estimating time-varying elasticities of Swiss gasoline demand. Energy Economics 34 1755–1762.

[9] F. Huneke, IJ. Henke, González JAB, Erdmann G. Optimization of hybrid off- grid energy systems by linear programming. Energy Sustainability Soc 2012; 2 (7):1–19.

[10] I. El Kafazi, R. Bannari, A. Abouabdellah. Management of energy production. *ARPN Journal of Engineering and Applied Sciences.* ISSN 1819-6608. VOL. 11, NO. 17, SEPTEMBER 2016.

[11] I. El Kafazi, R. Bannari, A. Abouabdellah. Modeling and Forecasting Energy Demand.4th International Renewable and Sustainable Energy Conference (IRSEC'16 November 14-17, 2016), p. 746 - 750.

[12] I. El Kafazi, R. Bannari, A. Abouabdellah, My O. ABOUTAFAIL and Josep M. Guerrero. Modeling and Forecasting Energy Demand.5th International Renewable and Sustainable Energy Conference (IRSEC'17December 04-07, 2017). (ACCEPTED).

[13] I. El Kafazi, R. Bannari, A. El Bouzekri El Idrissi, N. Hmina and T. Dragicevic. Renewable energies: modeling and optimization of production cost. 4th International Conference on Energy and Environment Research, ICEER 2017, 17-20 July 2017, Porto, Portugal.

[14] F. Egelioglu, A. A. Mohamad, H. Guven. Economic variables and electricity consumption in Northern Cyprus. Energy 2001; 26 (4):355–62.

[15] M. Bessec, J. Fouquau. The non-linear link between electricity consumption and temperature in Europe: a threshold panel approach. Energy Economics 2008; 30 (5):2705–21.

[16] J. L. Harris, L. Lon-Mu. Dynamic structural analysis and forecasting of residential electricity consumption. Forecasting 1993; 9:437–55.

[17] C. C. Lee, C. P. Chang. The impact of energy consumption on economic growth: evidence from linear and nonlinear models in Taiwan. Energy 2007; 32(12):2282–94.

[18] Global trends in renewable energy investment, report shows; Frankfurt School, FS-UNEP Collaborating.

IOP Conf. Series: Earth and Environmental Science **154** (2018) 012007 doi:10.1088/1755-1315/154/1/012007

Designing a dormitory with emphasis on renewable energy

F. Daneshvar Tarigh[1], A. Daneshvar Tarigh[2], and F. Habib[3]
[1] Department of Art and Architecture, Science and Research Branch, Islamic Azad University, Tehran, Iran
[2] Department of mechanical engineering Karlsruhe institute of technology (KIT), 76131 Karlsruhe, Germany
[3] Department of Art and Architecture, Science and Research Branch, Islamic Azad University, Tehran, Iran

Abstract. The majority of universities provides on- and off-campus residential quarters for students during their studies which enables them to keep connected to other students and focus on their studies usually with a small amount of money. The manner of designing a dormitory has a direct impact on the performance of the students and therefore requires a lot of attention. This includes but not limited to a mostly independent and private quiet room maintaining good indoor air quality through adequate ventilation and air conditioning. Undoubtedly, the most important aspect of such a place is saving energy in a way that does not influence the quality of student's life. The type of usage of such buildings causes different presence time and different ideas about the lights and temperature's set point. In this paper, we will discuss aspects of designing a dormitory as well as optimization of occupants comfort and energy efficiency using renewable energies such as solar energy to produce electricity, wind energy for natural ventilation and above all using architectural techniques to lower the energy consumption.

1. Introduction

Over the past few decades, the boom of energy demand in the Building sector led to climate change and global warming. Student dormitory is a very important facility which have to be provided by university. University should not only provide accommodation but also other supporting facilities [1]. It has been estimated that buildings are responsible for 40% of energy consumption and 36% of CO2 emissions in the Europe [2], [3]. To meet our goals, developing more eco-friendly sustainable design concepts offers the greatest opportunity to reduce the negative impacts on our environment.

2. Characteristics and physical location

The project is in a land inside the Islamic Azad University of Tehran, Science & Research Campus. In northern part of Tehran, capital of Iran, which is in Hot-dry with cold winters climate. [4] The average annual temperature is 17.1°C. It is located 35.69 latitude and 51.42 longitude and it is situated at elevation (average) 1178 meters above sea level.

The main feature of the vernacular buildings of hot-dry with cold winters is the provision of more open spaces. Central courtyard is very suitable for summer to keep the coolness and humidity of night and give refreshment during the summer days. It is also suitable for winter to protect the rooms from winter winds [5].

3. Design standards

According to the university's standard, we have considered at least 12 m2 per student as separate bed rooms, in six-room flats, each contains a kitchen, dining area and provide one separate WC with small wash hand basin and shower

IOP Conf. Series: Earth and Environmental Science **154** (2018) 012007 doi:10.1088/1755-1315/154/1/012007

- WC Minimum 1200mm x 1000mm.
- Wash hand basin (typically 500mm x 420mm) with cupboard under.
- Minimum 800mm x 800mm shower enclosure.

A full, typical student bedroom, including all interior finishes, trim, furnishings, windows, doors, heating, cooling and fresh air delivery units [6].
Martin Heilweil [7] prepared a listing that embraces most of the conditions cited by others. In somewhat abbreviated form they are as follows:

- Provision for either solitary study, or study with a minimum of others
- A place exclusively devoted to study
- Is free from distractions and movements of others
- Is free from noises from such sources as telephones, plumbing, and typing
- Is equipped with personal control of heat, light, and ventilation

4. Building envelope

The building envelope's impact on energy consumption should not be underestimated. Globally, space heating and cooling account for over one-third of all energy consumed in buildings, rising to as much as 50% in cold climates and over 60% in the residential sub-sector in cold climate countries. Therefore the building envelop should have the following standards:

- High levels of insulation in walls, roofs, and floors, to reduce heat losses.
- High-performance windows, with low thermal transmittance for the entire assembly (including frames and edge seals) and climate-appropriate solar heat gain coefficients (SHGC).
- Highly reflective surfaces, including both white and "cool-colored" roofs and walls, with glare minimized.
- Properly sealed structures to ensure low air infiltration rates, with controlled ventilation for fresh air
- Minimization of thermal bridges (components that easily conduct heat), such as high thermal conductive fasteners and structural members, while managing moisture concerns within integrated building components and materials. [8]

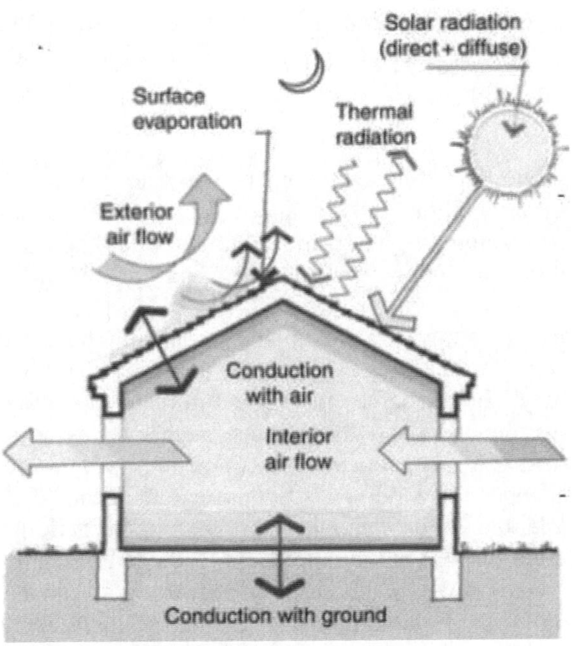

Figure 1. Paths of energy exchange at the building microclimate scale Watson and Labs, 1993

4.1. Walls

The structural performance of the building materials also requires careful consideration to all of its component including walls. Wall is a structure that refers to an area, carries a load, or provides shelter and security. Recent studies show that one of the best material to be used in constructions is light expanded clay aggregate (Leca). They can be used as a primary material in the mix design of low density building blocks and panels. Expanded clay block (Leca blocks Figure 2) and panel walls deliver lightweight, yet, robust building construction solutions.

Leca blocks are prefabricated elements, made of lightweight concrete mixed with cement, sand and water, and produced in rotary kiln at about 1200°C. Using of Leca aggregate results in decreasing the concrete density. Due to its lightness and structure, the material has good thermal and acoustic insulation and fire-proofed stability. It is therefore used mainly in the construction of basement walls, floors, internal partition walls and ceilings. [9]

Figure 2. Leca block

Hassan et al. [9] have done a comparative investigation on U-value (thermal transmittance) of some common materials. The results show that Leca block wall has lower U-value than the timbre wall. Technically, Leca block contains air-filled pores with the air-filled clayey cells, which contribute to its insulation properties. A pure solid Leca block has not significantly large thermal insulation, but in combination with sufficiently thick insulation, the structure will have sufficient insulation to meet the requirement for passive houses. The insulation in Leca block consists of polyurethane (PUR) produced by foaming a blowing agent. With this formed gas in a closed cell system, the insulation capacity will be improved. [9]

4.2. Fenestration (windows)

Improved building design can also offer significant potential to reduce the demand for lighting in buildings, through building orientation and advanced fenestration technologies such as dynamic windows. With better use of natural lighting and adoption of highly efficient lamp technologies, buildings energy consumption for lighting could be reduced by 40% in 2050 compared to current levels. [10]

While it is difficult to estimate, windows are most likely responsible for 5% to 10% of the total energy consumed in buildings in OECD countries. They fulfil multiple functions, including access to the building, outlook, entry of daylight and safety egress, and in many cases they provide ventilation as a fresh air inlet. In most cases, windows should let in as much light as possible, but this needs to be balanced by the fact that heat gain in summer needs to be minimized, while in winter it should be maximized. One of the challenges for windows is to optimize the heat flow depending on the season. If outdoor temperature is cold, the window should retain heat in the building, minimize losses and let in as much solar radiation as possible. On the other hand, if the temperature inside the building is too hot and cooling is needed, the windows should keep out heat from the sun and provide opportunities to shed heat from the building [10]. Three components define the thermal transmittance of a window (UW): the glass panes, the frame (fixed or operable), and the spacer between panes (multi glazed windows) [11].

IOP Conf. Series: Earth and Environmental Science **154** (2018) 012007 doi:10.1088/1755-1315/154/1/012007

4.2.1. Glass panes

Common glass panes are transparent to solar radiation in the range of wavelengths from ultraviolet to the near infrared (from 0.3 to 2.5lm), with a maximum peak in the visible range (about 42 % of the solar energy is emitted in the range of0.38–0.74lm wavelengths) [11]. An effective way to control the temperature gain through glass panes is to have smart glazing windows. Smart glazing can be divided into two major categories: non-electrically activated (photochromic and thermo-chromic) and electrically activated types. The most popular and most complicated is electro-chromic. The electrically activated devices have the advantage of automatic control but they are very expensive and they use electricity (approx. 5w/m2) Figure 3.

Figure 3. ELECTROCHROMIC Glass

Thermo-chromic laminated glazing (TLG) is a "smart" glazing that regulates the entry of light on a programmable and automatic basis. No electrical power, or other power is needed to operate the thermo-chromic system. Use of thermo-chromic material allows the creation of "intelligent" windows and facades Figure 4. These techniques allow building, with extensive glazing, to adapt dynamically and continuously to ever changing climatic conditions – and thereby minimize energy requirements. Thermo-chromic materials are affordable and have a service life of over dozen years. [12]

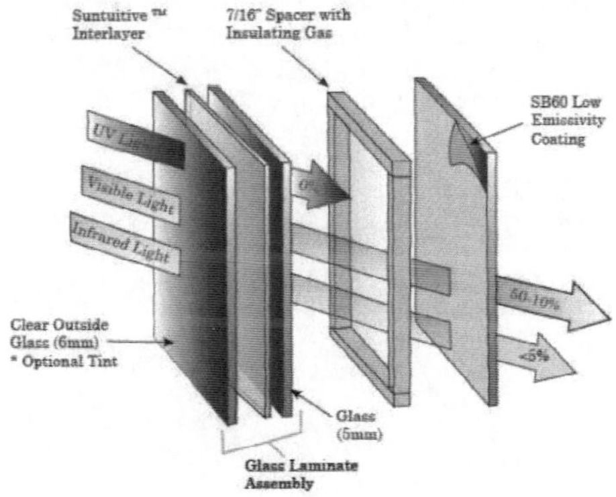

Figure 4. Sunlight-Responsive Thermochromic Window Systems

Source: OFFICE of ENERGY EFFICIENCY & RENEWABLE ENERGY

4.2.2. Frames

The frames can be classified according to the system of beaten into three categories:

- Windows with single beaten
- Windows with double or triple stop beaten
- Window frame with open joints.

The frames with single stop profile have a simple single seal, which has to guarantee air and water tightness; therefore, it is generally not reliable, especially in the presence of high external pressure, when the wing tends to inflect the frame itself causing the detachment of the gasket. At this purpose, the sealing of windows with double or triple stops should be preferred. On the other hand the frame material plays a key role in saving energy, the best type for this purpose is unplasticized polyvinyl chloride (UPVC). Multi–Chamber UPVC contributes to excellent thermal and sound insulation Figure 5.

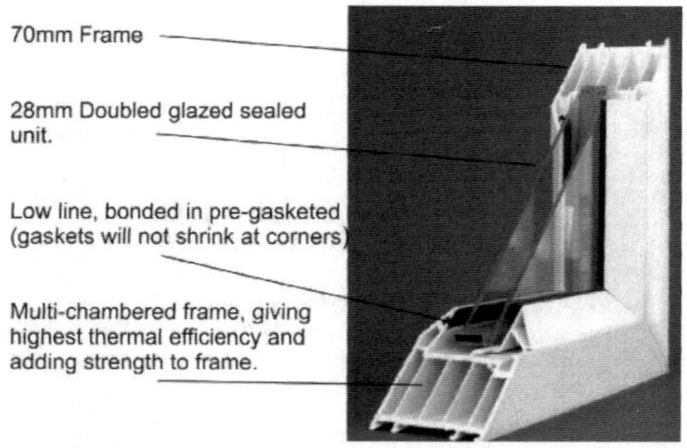

Figure 5. Multi–Chamber UPVC window frame www.diypvcwindows.com

5. Automatic window rolling shutters

A rolling shutter device as a component of a window system provides protection from direct sun and overheating in summer, and thus, reduces the cooling loads for the building and reduces the heating loads by preventing direct exposure of window to cold wind in winter, provides protection from glare Figure 6. Most commercially available windows cause heat loss in winter and heat gain (e.g., solar heat gain) in summer, and are usually not integrated with the other building systems. Blinds installed in these windows are occasionally part of the day lighting and thermal systems Figure 7. [13]

Figure 6. Window rolling shutters www.dustymars.net

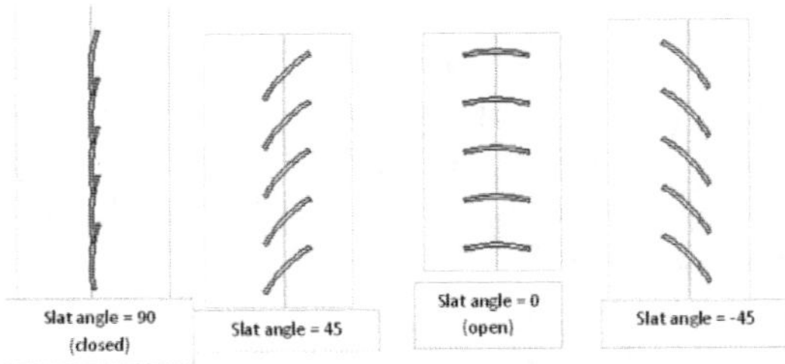

Figure 7. Louvered Blind [14]

5.1. Overhangs

Roof overhangs have several important functions: they can protect exterior doors, windows, and siding from rain; they can shade windows when solar heat gain is undesirable; and they can help keep basements and crawl spaces dry.

Overhang should be designed in a way that provides shade well into summer and full sun through the winter [15]. Donald Watson et al. [16] believed that size of overhang can approximately be calculated according to (1).

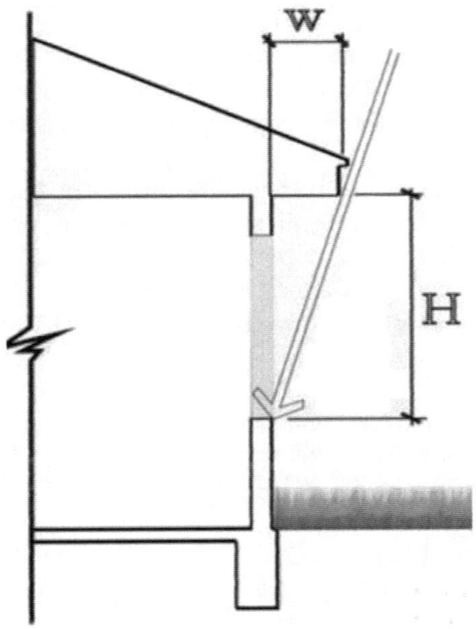

Figure 8. Overhang [13]

$$W=H/SLF \quad (1)$$

Where W is size of overhang, H is the height from the beneath of the overhang to the bottom of the window and SLF is coefficient which is obtained according to the (see 错误!未找到引用源。).

Table 1. Calculating the size of overhang according to latitude [13]

Window directions	Latitude°						
	25	30	35	40	45	50	55
East	0.8	0.8	0.8	0.8	0.8	0.8	0.8
South East	1/9	1.6	1.4	1.3	1.1	1.0	0.9
South	10/1	5.4	3.6	2.6	2.0	1.7	1.4
South West	1/9	1.6	1.4	1.3	1.1	1.0	0.8
West	0.8	0.8	0.8	0.8	0.8	0.8	0.8

6. Lighting

Light is the stimulus that influences most the human perception, but also the psychophysical wellbeing of the individual in everyday life [17]. On the other hand lighting accounts for a substantial amount of a building's energy budget. Estimates vary in the literature, with percentages from 20% to 40% being reported [18]. Hence the best solution which is energy efficient and durable is Light-Emitting Diode (LED). While recommended for studying is 500 lux/m2, a 50 watt LED lamp produces 6000 lux which is enough for 12 m2.
Meanwhile using motion sensor to turn off unnecessary lights would be beneficial.

7. Energy-electricity generation

While the project has a land of 13600 m2 and considering 3600 m2 for the installation of outdoor equipment (HVAC etc..), we can have 10000 m2 to be used to install sustainable energy generation devices such as Photovoltaic (PV) or solar cells.
PV systems behave in an extraordinary and useful way: They react to light by transforming part of it into electricity. Moreover this conversion is novel and unique, since photovoltaic [19]:

- Have no moving parts (in the classical mechanical sense) to wear out.
- Contain no fluids or gases (except in hybrid systems) that can leak out, as do some solar-thermal systems
- Consume no fuel to operate
- Have a rapid response, achieving full output instantly
- Produce no pollution while producing electricity
- Can be made from silicon, the second most abundant element in the earth's crust
- Have a relatively high conversion efficiency giving the highest overall conversion efficiency from sunlight to electricity yet measured
- Have wide power-handling capabilities, from microwatts to megawatts
- Have a high power-to-weight ratio making them suitable for roof application
- Are amenable to on-site installation, i.e., decentralized or dispersed power

The two major type of solar cells are mono-crystalline and polycrystalline which mono-crystalline are more expensive but Shuvajit Roy et al. [20] proves that they are the most efficient in comparison to others.

Table 2. Solar cell comparison www.solarcells.com

	Mono-crystalline	Poly-crystalline	Amorphous (Thin Film)	CdTe (Thin Film)	CIS/CIGS (Thin Film)
Typical Module Efficiency	17-22%	14-18%	7-9%	10-14%	11-13%
Best research efficiency	25.0%	20.4%	13.4%	19.6%	20.8%
Area required for 1 KWp	6-9m^2	8-9m^2	13-20m^2	11-13m^2	9-11m^2
Lowest price/watt (module only)	\$0.75/watt	\$0.55/watt	\$0.69/watt	\$0.59/watt	
Temp Resistance	10-15% perf. drop in high temps	More tolerant than mono	Tolerates extreme heat	Tolerates extreme heat	Tolerates extreme heat

Table shows that using 10000 m2 of the project (including remaining land and building roofs) can roughly obtain 1.5MW of sustainable power which is considerable amount.

8. Cooling heating and ventilation

Thermal comfort is important for both health and performance of occupants. In 1970s, Fanger proposed the PMV (Predicted Mean Vote) model based on heat balance of human body, and this model indicated that people's thermal sensation can be estimated by measuring air temperature, relative humidity, air velocity, mean radiant temperature (MRT), clothes insulation and metabolic rate.

Nowadays, PMV model has been adopted by ASHRAE Standard and ISO Standard. However, PMV model is also challenged by the theory of adaptive comfort which was developed since 1970s, because PMV model overestimates or underestimates actual sensation of occupants in the environment deviating from moderate condition, especially in naturally ventilated buildings [21].
Space heating, water heating and space cooling account for nearly 55% of global buildings energy use and represent the single largest opportunity to reduce buildings energy consumption in most regions of the world. A systems approach, including integration of heating and cooling needs with improved building envelopes, is necessary to achieve higher energy efficiencies and a low-carbon heating and cooling supply [10].
Geothermal technology (heat pump) taps the stored energy in the earth's crust. These systems use the earth's relatively constant temperature to provide heating, cooling, hot and cold water for homes or commercial buildings. For closed loop systems, water or an antifreeze solution is circulated through pipes buried beneath the earth's surface. During the winter, the fluid is heated from the earth and carries the heat through the system and into the building. During the summer, the system reverses the heat cycle in order to cool the building by pulling heat from the building; carrying it through the system and placing it in the ground [22].

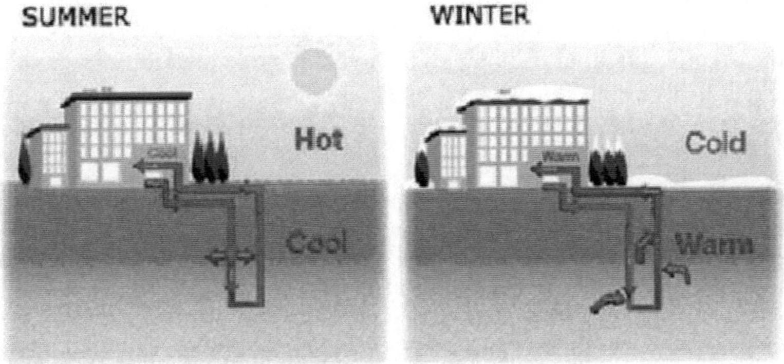

Figure 9. Geothermal Heating and Cooling Systems www.ozarker.org

A heat pump is a device that forces the movement of heat from a low-temperature medium to a higher temperature medium Figure 9. A GSHP transfers energy to and from a ground or water source to provide heating or cooling. In heating mode, the energy produced by this technology is considered partially renewable because solar and geothermal energy is mediated through the ground or water source. Depending on the generation source of electricity, the energy can be fully renewable. A GSHP system is typically composed of a ground loop (tubing that passes through a ground or water source, transferring energy to circulating fluid), a heat pump (a mechanical system that allows for the extraction of energy from the ground-loop fluid), and a heat distribution system (the system that distributes heat throughout a conditioned space).

8.1. Benefits of geoexchange
Geoexchange technology has several benefits, including:
Low Operating Cost - The efficiency of the heat pumps operating under moderate loop temperatures provides the basis for high efficiency and low operating cost. The cost to move energy around the building is also low, as heat pumps are placed at each space. There is no need to circulate large amounts of air around the building to transport energy, nor is there a need to reheat air to maintain comfort in certain areas of a building.
Simplicity – The distributed nature of the system makes it easy to understand. A heat pump located at each space will provide independent heating and cooling. The operation of one heat pump does not affect any other heat pump. Control simply requires turning the unit on or off in response to the area that needs heating or cooling.

IOP Conf. Series: Earth and Environmental Science **154** (2018) 012007 doi:10.1088/1755-1315/154/1/012007

Low Maintenance – The heat pump itself is a packaged unit no more complex than typical residential air conditioning equipment. The components are the same as those used for outdoor applications that have much wider operating ranges and exposure to the weather. Diagnosing problems has become easier due to the distributed nature of the system. Any problem is typically closely related to the equipment serving the particular space.
No Supplemental Heat Required – Heat pumps can meet all of the space loads, including ventilation loads. Ventilation air can be tempered by separate heat pumps and/or conditioned with heat recovery equipment.
No Required Exposed Outdoor Equipment – The ground heat exchanger is buried and the heat pumps are located inside the building. Vandalism, noise, and visual screen problems are eliminated. Designers do not have to supply space on the roof for equipment, making options such as standing seam metal roofs or large sloped roofs possible.
Low Environmental Impact – No fossil fuels need to be consumed on site. Pollution can be best mitigated at a central power plant where electricity is produced. As the efficiency of electricity production or renewable power generation increases, so does the environmental efficiency of the heat pump system.
Level Seasonal Electric Demand – With winter heat pump operation displacing fossil fuel use, and summer heat pump operation occurring at moderate, more efficient loop temperatures, the electric demand is more consistent throughout the year so the average price of electricity is reduced.
Longer Life Expectancy - Both the American Society of Heating Refrigerating and Air-Conditioning Engineers (ASHRAE) and the Electric Power Research Institute have concluded, based on independent research studies, that the appropriate service life value for ground source heat pump technology is 20 years or more. This benchmark is the current industry standard.

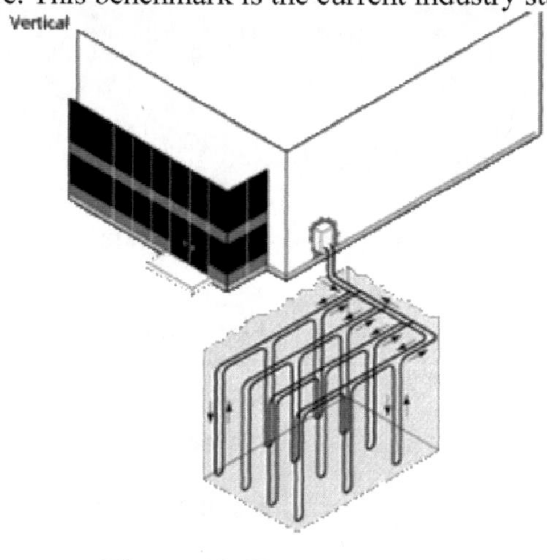

Figure 10. Heat pump [23]

8.2. Solar water heating

A solar hot water heater heats the same amount of water for a fraction of the cost Figure 11. A solar hot water heating system's performance is dependent on the intensity of the sun in its location. The cost of solar water heating from large plants is four to six times lower than that of heat from small plants [24]. Therefor it can be considered to assist the providing the running hot water of the dormitory. Solar water heating features parallel rows of transparent glass tubes Figure 12. Each tube contains a glass outer tube and metal absorber tube attached to a fin. The fin's coating absorbs solar energy but inhibits radiative heat loss. These collectors are used more frequently for commercial applications [25].

Figure 11. Solar water heating

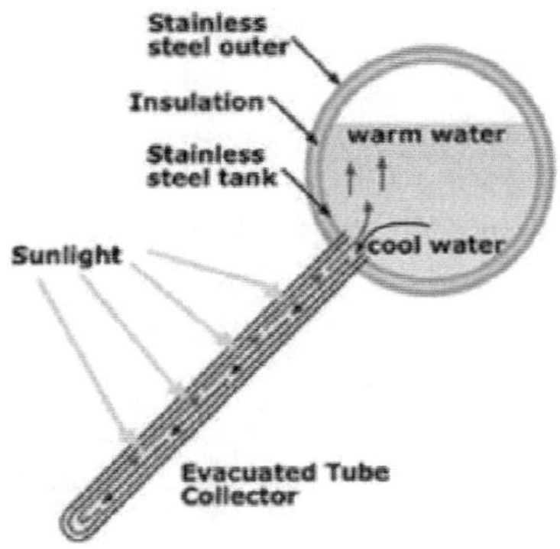

Figure 12. Solar water heating function [25]

8.3. Passive cooling techniques

Natural ventilation is a passive cooling method, can reduce building energy consumption, so by a high degree of attention. The study of the law of air flow and the influence on the thermal environment in the students' dormitory under the condition of natural ventilation is an important condition to ensure the environment of scientific design dormitory [26].

There are four design methods available for natural ventilation systems:

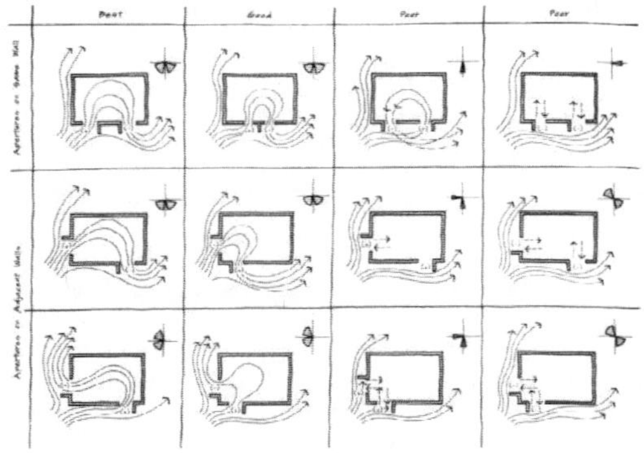

Figure 13. Relationships between building openings and airflow [23]

IOP Conf. Series: Earth and Environmental Science **154** (2018) 012007 doi:10.1088/1755-1315/154/1/012007

Cross flow (no corridor), the simplest natural ventilation system with no obstacles on either side of the prevailing wind (i.e. windows of similar size and geometry open on opposite sides of the building) Figure 14.

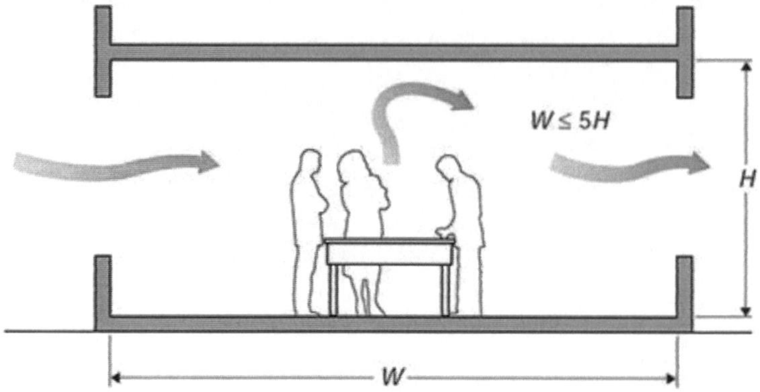

Figure 14. Cross flow ventilation www.abec.co.uk

Wind tower (wind catcher/wind extractor), the positive-pressure side of the wind tower acts as a wind catcher and the negative-pressure side of the wind tower acts as a wind extractor.

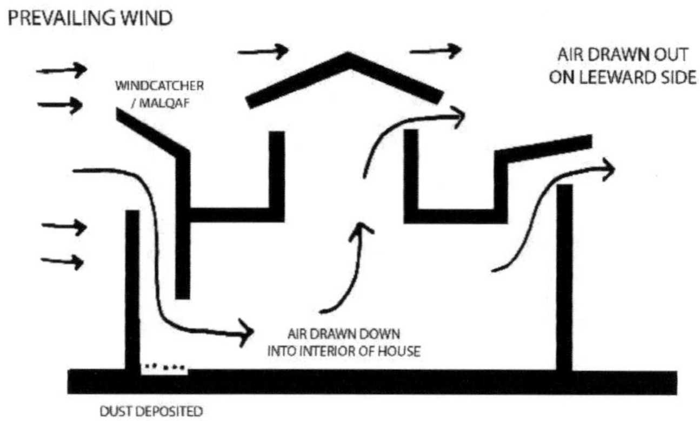

Figure 15. Wind tower ventilation / commons.wikimedia.org

Stack (or buoyancy), simple flue, a vertical stack from each room, without any interconnections goes through the roof; this allows for air movement based on density gradients

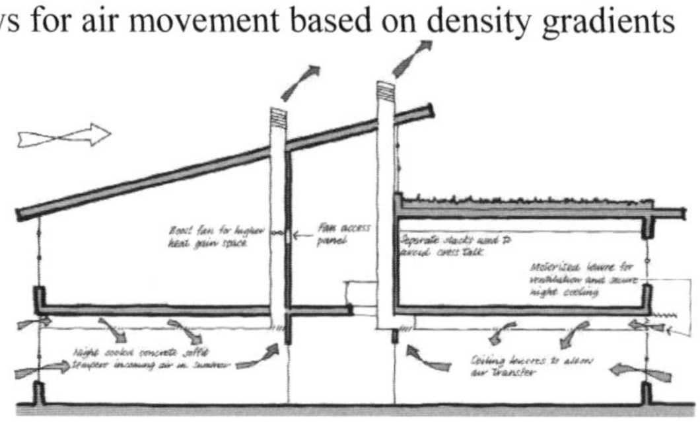

Figure 16. Stack ventilation www.passivent.com

Stack (or buoyancy), solar atrium, a large stack that heats due to solar radiant loading, which induces air movement due to density (temperature) differentials; without radiant loading, the atrium provides minimal ventilation **Figure 17**.

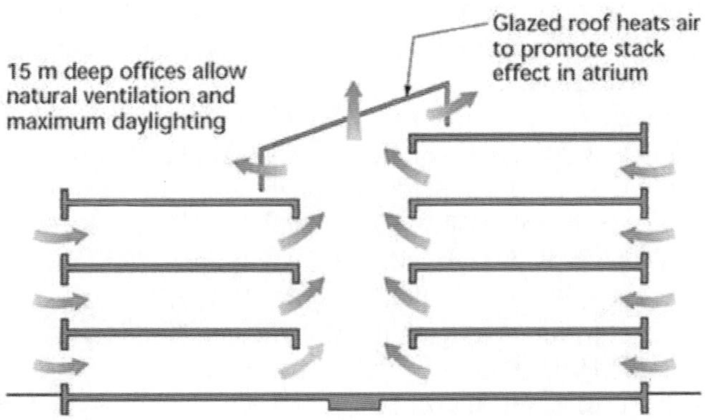

Figure 17. Stack ventilation www.passivent.com

9. Conclusion

This study present a functional method to design a dormitory on the bases of renewable energy. Building envelope including walls, windows, and shutters are chosen concerning the energy saving aspect. We chose LECA Blocks to be used in walls to have considerable insulation meeting the requirement for passive houses. Different parts of windows have been studied. Frames are designed to use Multi–Chamber UPVC which contributes to excellent thermal and sound insulation, Glass panes are thermo-chromic laminated smart glazing which controls the sun light. We recommend using Automatic window rolling shutters to prevent direct sun and overheating in summer. Roof overhangs used to shade windows when solar heat gain is undesirable. LED lamps have been recommended to decrease the amount of energy used. We applied Mono-crystalline solar cells to generate energy. An energy efficient cooling, Heating and Ventilation has been considered.

10. References

[1] C. Susilawati, "Student Dormitory Development Plan with Linear Programming Method," Pacific Rim Real Estate Society (PRRES), 2001.

[2] C. T. D. A. F. P. Marialuisa Baggio, "Sustainability of a Historical Building Renovation Design through the Application of LEED Rating System," Energy Procedia, vol. 113, pp. 382-389, 2017.

[3] "European Commission," 25 11 2017. [Online]. Available: https://ec.europa.eu/energy/en/topics/energy-efficiency/buildings.

[4] K. W. (1936). and G. Geiger, "Handbuch der Klimatologie," in Das Geographisca System der Klimate., Berlin, Verlog von Gebruder borntraeger, 1936, pp. 1-44.

[5] P. Pourvahidi and M. B. Ozdeniz, "Bioclimatic analysis of Iranian climate for energy conservation in architecture," Scientific Research and Essays, vol. Vol. 8(1), no. 4 January, pp. 6-16, 2013.

[6] B. o. R. o. t. U. S. o. Georgia, Baseline Design Standards For Student Housing - APPENDIX E, Atlanta: University System of Georgia, 2014.

[7] M. Heilweil, "THE INFLUENCE OF DORMITORY ARCHITECTURE ON RESIDENT BEHAVIOR," Environment and Behavior, vol. 5, no. 4, pp. 377-412, 1973.

[8] "TechnologyRoadmapEnergyEfficientBuildingEnvelopes," International Energy Agency, pp. 7-9, September 2013.

[9] P. J. Osama A. B. Hassan, "Lecablock, an Alternative Construction Material for the Exterior Walls of Passive House," Journal of Building Construction and Planning Research, vol. 2, pp. 96-108, 2014.

[10] "Transition to Sustainable Buildings Strategies and Opportunities to 2050," IEA/OECD ISBN: 978-92-64-20241-2, Paris, 2013.

[11] C. C. F. S. a. S. S. G. Cellai, "Transparent Building Envelope Windows and Shading Devices Typologies for Energy Efficiency Refurbishments," Building Refurbishment for Energy Performance. Green Energy and Technology. Springer, Cham, pp. 61-118, 2014.

[12] R. E. Arutjunjan, T. S. Markova, I. Y. Halopenen and I. K. Maksimov," Thermochromic Glazing for Zero Net Energy house," GDP Glass processing days, no. Technological University of Plant Polymers, pp. 299 - 301, 2003.
[13] J. H. Svetlana Olbina, "Daylighting and thermal performance of automated split-controlled blinds," Building and Environment, vol. 56, no. Elsevier, pp. 127-138, 2012.
[14] M. Y. C. K. R. H. R. M. S. V. D. Charlie Curcija, "Energy Savings from Window Attachments," Energy Efficiency & Renewable Energy, vol. October 2013, pp. 1-103, 2013.
[15] F.D.T.A.N.Alireza Daneshvar Tarigh, "A Survey of Energy-Efficient Passive Solar Houses," IPCBEE© IACSIT Press, Singapore, vol.27, pp.18-25, 2012.
[16] D. Watson, "Climatic Building Design:Energy-Efficient Building Principles and Practices," in Kenneth Labs, Mcgraw-Hill, 1993.
[17] G. G. E. T. Rosella Tomassoni, "Psychology of Light: How Light Influences the Health and Psyche," in University of Cassino and Southern Latio, Cassino, Italy, 2015.
[18] G. M. P. O. a. A. R. D. T Delaney, "Evaluation of energy-efficiency in lighting systems using sensor networks," in Proceedings of the 1st ACM Workshop on Embedded Sensing Systems for Energy-Efficiency in Buildings, ACM, New York, NY, USA, 2009.
[19] S. E. R. Institute, Basic Photovoltaic Principles and Methods, Washington, DC: Midwest Research Institute, 1982.
[20] R. S. S. T. F. SHUVAJIT ROY, "PERFORMANCE ANALYSIS OF MONO-CRYSTALLINE AND POLY-CRYSTALLINE SILICON SOLAR CELLS UNDER DIFFERENT CLIMATIC CONDITIONS: A COMPARATIVE STUDY," Dept. of Electrical & Electronic Engineering, BRAC University, Dhaka 1212, 2014.
[21] N. L. J. P. W. Z. Y. L. Yingdong He, "Field study on adaptive comfort in air conditioned dormitories of university with hot-humid climate in summer," Energy and Buildings, vol. 119, pp. 1-12, 2016.
[22] C. College, "BHT Dorm Project," Engineering 333, 2010.
[23] M. Karr, "Ground-Source Variable Refrigerant Flow Heat Pumps," Energy Engineer, Washington State University Extension Energy Program, Washington, 2011.
[24] "Ramboll Group," Ramboll, 2017. [Online]. Available: http://www.ramboll.com/.
[25] P. Christopher A. Homola, "Solar Domestic Hot Water Heating Systems Design, Installation and Maintenance," [Online]. Available: http://www.asse-plumbing.org. [Accessed 25 11 2017].
[26] L. J. Dong Yanga, "Simulation study on the Natural Ventilation of College Student' Dormitory," Procedia Engineering, vol. 205, no. 10th International Symposiumon Heating, Ventilation and Air Conditioning, pp. 1279-1285, 2017.

IOP Conf. Series: Earth and Environmental Science **154** (2018) 012008 doi:10.1088/1755-1315/154/1/012008

Comparative analysis of heat dissipation panels for a hybrid cooling system integrated in buildings

A Zuazua-Ros[1], JC Ramos[2], C Martín-Gómez[1], Tomás Gómez-Acebo[2] and A Pisano[2]

[1]Construction, Building Services and Structures Department, Universidad de Navarra, 31180, Pamplona, Spain
[2]TECNUN School of Engineering, Universidad de Navarra, 20018, San Sebastián, Spain

Abstract. The use of cooling panels as heat dissipation elements integrated in buildings has been previously investigated by the authors. Those elements would be connected to the condenser and would dissipate the heat in a passive form. Following the research, this study analyses and compares the thermal performance of two heat dissipation panels as part of a hybrid cooling system. Both panels were experimentally tested under different variables, thus having nine scenarios for each panel. Additionally, an already validated model was applied. The empirical results show a considerable difference between the cooling capacity among them, doubling the daily average ratio in one scenario. The heat dissipation ratios vary between 106 and 227 W/m^2 in the first case and 140 and 413 W/m^2 in the second. Regarding the model applicability, the average error for each panel was 4.0% and 8.5%. The bond between the metal sheet and the pipes of the panels has proven to be the main parameter to assure the highest heat dissipation potential of each panel.

1. Introduction

Energy demand for cooling is expected to increase in the residential, industrial and service sectors in the next 50 years [1]. Besides, cooling represents a significant use of energy globally and a meaningful driver of peak electricity demand. A recent study about the future of cooling in buildings concludes underlining the importance of the development of alternative cooling dissipation technologies based on the use of low temperature environmental sinks [2].

In this context, passive cooling systems are those techniques that use natural driving forces to circulate fluid, while active systems require external mechanical power to operate. Thus, hybrid systems are the combination of both, the use of natural driving forces and the need of mechanical power [3]. The components analysed in this study are part of a hybrid cooling system solution.

The origin of the solution derives from the attempt to propose an alternative to evaporative cooling systems, specifically those that include cooling towers [4]. Cooling towers are heat dissipation equipment that push the waste heat generated in cooling systems into the atmosphere. All cooling towers, to greater or lesser degree, have the following drawbacks: (a) High water use (it increases considerably in open circuit systems). In addition, several chemical products must be used for fluid treatment. (b) Legionella. It is widely known that cooling towers represent a source of Legionnaires' disease outbreaks [5]. (c) Maintenance requirements. Tower manufacturers specify the operation and maintenance manuals. Additionally, ASHRAE recommends an inspection and maintenance schedule [6]. Since the maintenance must be strict in a tower or evaporative condenser, it involves high costs in terms of personnel and products. (d) Noise and vibrations. These are mostly generated by fans. (e)

Operating cost. The items that make the greatest contribution to operating costs are water, electricity (pumps and fans), chemicals (to maintain water quality) and operators. All these components made necessary to develop an alternative system that would eliminate or reduce the environmental and economic impact of cooling towers.

In this context, an alternative application of a heat dissipation solution was proposed. The solution deals with the design of heat dissipation surfaces integrated in building façades. Those surfaces would dissipate the heat generated in cooling cycles and seek to reduce the energy consumption for cooling in buildings. In this way, the drawbacks of evaporative cooling systems would be reduced or eliminated. The diagram of the system is shown in Figure 1.

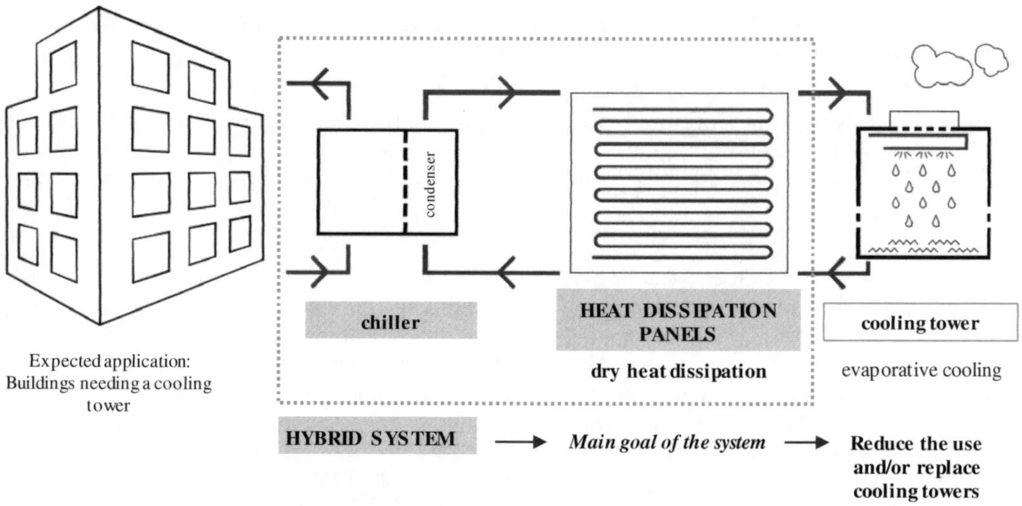

Figure 1. Schematic diagram of the solution.

The heat sink used is the ambient temperature and the heat transfer method is by radiation and convection. The system resembles to night-time radiation systems, where the sky temperature is the heat sink. However, the operation temperatures are higher in this study, allowing the system to operate 24 hours. There are other studies based on radiative surfaces as heat dissipation method. For example, in the study by Erell [7] has been calculated radiative cooling using unglazed flat-plate solar collectors, connected to a storage tank and operating during night. Other recent studies have used flat plate collectors on rooftops, saving the 10% of the power consumption for cooling purpose during overnight exposure in different cities of Malaysia [8]. In the study of Cui et al., the cooling capacity of a radiative panel has been analysed depending on the tilted angle, where the cooling potential decreases considerably in vertical position [9]. There are studies that consider the integration of the cooling system in the construction, as [10], where the experimental results showed an average cooling capacity that reaches 87 W/m^2 in Tianjin, China. The panel in the present study has been supposed to work in the external layer of the façade of a building, in order to be exposed to the highest natural ventilation possible. A similar envelope design was made by [11] in Japan, which is installed in an already existing building.

In this case the proposal has three main characteristics that must be addressed: (1) the heat sink used is the ambient air temperature and surroundings, (2) the heat transfer method is by radiation and convection and (3) the final aim of the solution is to integrate the panels in a vertical position in a building envelope. The preliminary results has shown a potential energy savings between 11 and 33 % depending on the operation conditions of the cooling panels [12].

In this document, a comparison between two options for the cooling panel is made in order to analyse on the one hand, the heat dissipation capacity difference and on the other, the applicability of the mathematical model previously validated with the first panel. Besides, a brief evaluation of the parameters that affect the performance is made and an optimized panel option is studied.

IOP Conf. Series: Earth and Environmental Science **154** (2018) 012008 doi:10.1088/1755-1315/154/1/012008

2. Experimental setup and model description

Two heat dissipation panels will be analysed. The first one, hereinafter referred as cooling panel, was assess to validate the mathematical model presented in [13]. Initially, it was commercialised for chilled indoor ceilings (model WK-D-UM by TROX) but it was tested outdoors in a vertical position, see Figure 2. The second panel, hereinafter solar collector, consist on a solar thermal collector obtained from the company Junkers (model FKC-2), in which the glass cover has been removed, Figure 3.

Figure 2. Cooling panel TROX.

Figure 3. Solar collector by Junkers.

In both cases the experimental rig is completed with a closed hydraulic circuit that includes a chiller, a thermostatic bath, a flow meter with a precision of 1.26% of full scale, a needle valve and the connection lines, as shown in Figure 4. Regarding the temperature control and measurement, twenty-two T-type thermocouples with a precision of ±0.5°C were installed to measure the temperature in different part of the panel. Moreover, other thermocouples were installed to measure the temperatures of the ambient air (T_{amb}) and the inlet (T_{in}) and the outlet (T_{out}) temperatures. Regarding the heater, the temperature (T_h) was calculated directly by bath's sensor. The prototype was set up in San Sebastian (in the north of Spain) and the experiments were carried out during June and July 2016 for the cooling panel and May and June 2017 for the solar collector.

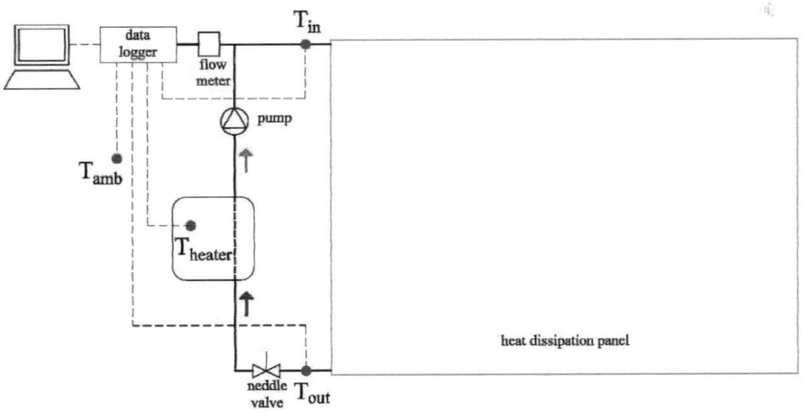

Figure 4. Experimental setup schematic.

The thermal conductivity of the union between the pipes and the metal sheet, expressed as C_s, is one of the key parameters for obtaining a high dissipation ratio. It is calculated from the steady state values taken from experiments carried out indoors, with a constant ambient temperature. The value obtained in each panel differs considerably, since it is 1.5 W/m K for the cooling panel, where the union between the coil and the support plate is press-clipped, and 25 W/m K for the solar collector, where the coil is welded. In Table 1 the main physical characteristics of each panel are shown.

The panels were placed in a vertical position (tilt angle 90°), north oriented, avoiding as much as possible the solar beam radiation. Nine different scenarios were analysed, with three inlet temperatures

IOP Conf. Series: Earth and Environmental Science **154** (2018) 012008 doi:10.1088/1755-1315/154/1/012008

(35, 40 and 45ºC) and three fluid flowrates (0.5, 1 and 1.5 l/min). Each test was named to reflect the scenario; the names consist of the inlet temperature followed by the fluid flow rate. For example, test T35q05 has an inlet temperature of 35°C and a volumetric flowrate 0.5 l/min.

Table 1. Main characteristics of the cooling panel and the solar collector.

Parameter	Symbols	Unit	Cooling panel	Solar collector
Support plate				
Area	A	m^2	2.40	2.18
Plate thickness	t	mm	1.5	1
Plate height	L	m	1	1.09
Solar absorptivity	α	-	0.9	0.9
Emissivity	ε	-	0.9	0.99
Pipes				
Distance between tubes	d_t	mm	70	90
External tube diameter	D_{ext}	mm	12.8	8
Internal tube diameters	D_{int}	mm	11.8	7
Number of coils	-	-	14	22

The previously developed model, consists on the calculation of the outlet temperature of the water after dissipating the heat by means of natural convention and radiation to the ambient air and the surroundings. In [13], the whole development of the calculations is presented. However, for this paper, just the main equations that affect the parameters that will be analysed are shown.
The fluid outlet temperature is the unknown parameter that is required to calculate the total heat dissipation of the panel. The calculation is expressed in the energy balance equation (1). This equation calculates the net cooling potential of the panel.

$$\dot{Q} = \dot{m} \cdot \bar{c}_p \cdot (T_{in} - T_{out}) \tag{1}$$

In order to calculate the fluid outlet temperature, equation (2) is used, from Duffie and Beckman [14]. This equation calculates the fluid outlet temperature as a function of the rest of the parameters of the cooling panel.

$$\frac{T_{out} - T_{amb} - \dot{S}/U_L}{T_{in} - T_{amb} - \dot{S}/U_L} = \exp\left(-\frac{A_p \cdot U_L \cdot F'}{\dot{m} \cdot c_p}\right) \tag{2}$$

The panel efficiency factor, F', represents the relation between the real energy loss of the panel and the energy loss that would be obtained if the panel achieved the same temperature as the fluid. F' is given by equation (3).

$$F' = (1/U_L)\left\{d\left[\frac{1}{U_L[(d_t - D_{ext})\eta_f + D_{ext}]} + \frac{1}{h_{\text{int}}\pi D_{\text{int}}} + \frac{1}{C_s}\right]\right\}^{-1} \tag{3}$$

where d is the distance between the tubes, D_{ext} is the external diameter of the tube and η_f is the fin efficiency factor. In this case the fin is represented by the perforated plate, and its efficiency is given by equations (4) and (5).

$$\eta_f = \frac{\tanh[m(d - D_{ext})/2]}{m(d - D_{ext})/2} \tag{4}$$

$$m = \sqrt{U_L/k_p t} \tag{5}$$

To evaluate the performance of the cooling panel, efficiency was defined as the ratio of the real heat transfer to the maximum heat transfer if the outlet temperature was equal to the ambient temperature, equation (6).

$$\eta = \frac{T_{in} - T_{out}}{T_{in} - T_{amb}} \tag{6}$$

The mean temperature of the panel, which is needed to calculate the overall thermal losses coefficient, U_L, is taken as the mean temperature of the inlet and the ambient temperatures, which is the correlation found from the initial experiments. Besides, due to the low incidence of the solar radiation shown in the experiments, the S is composed only by the direct radiation, and the reflected and diffuse components of the solar radiation are disregarded. The values of the solar radiation are taken from month average values based on the Spanish regulation CTE [15].

3. Experimental results and simulation

The analysis of the results is divided in three parts, first the experimental data comparison, then the applicability of the mathematical model in both heat dissipation panels and finally the analysis of an improved proposal.

3.1. Experimental results

From each scenario, at least 24 hours of test data are obtained, recording the values every minute. As it is predicted, the ambient temperature affects directly the outlet temperature, when the ambient temperature increases, the outlet temperature increases as well, reducing the heat dissipation capacity.

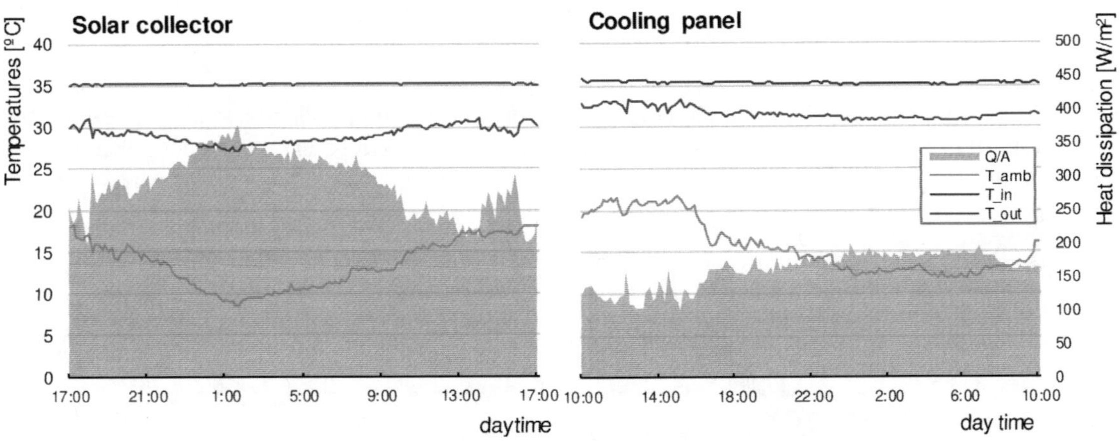

Figure 5. Experimental results comparison of the scenario T35q15.

The graphs from Figure 5 show the typical days data, when the outlet temperature decreases the dissipated heat increases. The solar collector dissipates more heat (corresponds to the grey shading, in W/m^2) than the cooling panel under the same variable conditions of inlet temperature and fluid flow. The ambient temperature averages for each 24-hour test are 13.5 °C for the solar collector and 15.6 °C for the cooling panel, which partially could explain the difference in the heat dissipation, however, when both ambient temperatures are around 15 °C, the heat dissipation is always clearly higher with the solar collector.

Regarding the dissipated heat ratio, Table 1 gathers the data about the average dissipation and the standard deviation of each scenario, along with the efficiency and the ambient temperature average of each panel. Comparing the heat dissipation ratio, the solar collector always dissipates more, even when the ambient average temperature is higher. In addition, the dissipated heat is always higher when the fluid flow increases in every case. The difference between the panels reaches a maximum of 106 %

in the scenario T35q15, which is the case of Figure 4. In average, considering the nine scenarios, the difference on the heat dissipation is 75 %. The performance values also are always higher in the results obtained with the solar collector. For both panels, when the fluid flow rate increases, the performance decreases.

The heat dissipation difference between the panels is mainly due to the efficiency factor value, F' (equation 3). The efficiency factor mean value for the cooling panel is 0.56, however, the solar collector has an efficiency factor of 0.91. This difference is caused by the thermal conductivity of the union between the pipes and the metal sheet, C_s, which differs considerably among both panels, as explained in section 2.

Table 2. Summary of the outdoor results.

Scenarios		Cooling panel			Solar collector			Mean heat dissipation difference
T_{in} [°C]	q [l/min]	T_{amb} [°C]	Heat dissipation [W/m²]	η	T_{amb} [°C]	Heat dissipation [W/m²]	η	
35	0.5	17.9	106 ± 22	47%	19.1	140 ± 25	71%	32%
	1	20.8	109 ± 16	25%	21.1	174 ± 26	36%	59%
	1.5	18.7	141 ± 24	19%	13.6	291 ± 45	28%	106%
40	0.5	19.7	139 ± 21	49%	16.3	224 ± 27	58%	61%
	1	19.1	159 ± 20	27%	17.9	293 ± 19	41%	85%
	1.5	20.1	171 ± 17	19%	19.0	336 ± 71	31%	97%
45	0.5	18.1	174 ± 15	46%	21.1	247 ± 35	65%	42%
	1	19.6	195 ± 21	27%	20.6	326 ± 86	41%	67%
	1.5	17.7	227 ± 35	19%	17.1	413 ± 28	32%	82%

3.2. Model applicability

The model developed for the cooling panel is applied also for the solar collector to calculate the outlet temperature in each scenario. The graphs in Figure 5 are an example of the results achieved with the model fitting.

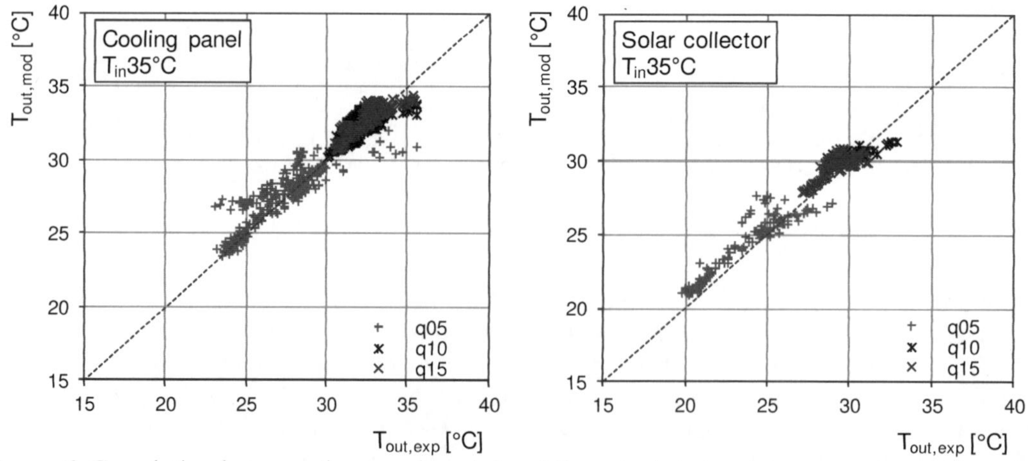

Figure 6. Correlation between the experimental and the calculated outlet temperature with an inlet temperature of 35 °C.

These graphs are used with the aim of enabling an initial reading of the error magnitude. Both variables are related and the deviation to the ideal scenario can be checked. The outlet temperature results in the cooling panel are higher than the outcomes in the collector. The graphs also show that the model generally gives a higher temperature than the reality.

Table 3. Daily total dissipated heat comparison between the experimental and model results.

scenarios		Cooling panel			Solar collector		
T_{in}	q	Total dissipated heat [kWh/m^2·day]		Difference	Total dissipated heat [kWh/m^2·day]		Difference
[°C]	[l/min]	experimental	model	[%]	experimental	model	[%]
35	**0.5**	2.55	2.46	-3.5	3.35	3.16	-5.8
	1.0	2.62	2.59	-1.3	4.18	4.22	1.0
	1.5	3.39	3.17	-6.4	7.04	6.61	-8.0
40	**0.5**	3.33	3.08	-7.3	5.37	5.17	-3.7
	1.0	3.81	3.70	-2.8	7.03	6.17	-12.3
	1.5	4.10	3.89	-5.0	8.07	6.67	-17.4
45	**0.5**	4.11	3.99	-2.8	5.93	5.55	-6.4
	1.0	4.67	4.52	-3.2	7.83	7.03	-10.2
	1.5	5.46	5.29	-3.1	9.92	8.55	-13.8

For Table 3, the periods where the system is not dissipating because of direct solar radiation are not considered. The average absolute difference in the case of the cooling panel is 4.0 %, however, for the solar collector, the difference increases to 8.5 %. Thus, the model developed and validated in [13], is more adjusted to the cooling panel than the collector. In the case of the solar collector, there are three scenarios where the error value exceeds 12 %. It may be addressed that in general the model shows dissipation values lower than the experimental results, except T35q10 in the solar collector.

3.3. F' affection analysis

As explained, the efficiency factor value in the case of the cooling panel is very low due to the poor bond between the metal sheet and pipes. The thermal conductivity of the bond is increased to 25 W/mK and a simulation is run to evaluate how much the heat dissipation would increase. The F' rises from 0.56 to 0.93, which influences in the heat dissipation average increase between 29 and 57 % comparing the experimental outcomes to the new calculated results. Moreover, the efficiencies shown in Table 2 for the cooling panel also increase up to 66 % for 0.5 l/min fluid flow rates, 40 % for 1.0 l/min and 29 % for 1.5l/min. This confirms the importance of the bond design to maximize the cooling potential of configuration type of panels.

4. Conclusions

This paper analyses and compares two heat dissipation panels (named cooling panel and solar collector) which will be components of a hybrid cooling system. The experimental study is made under the variation of two parameters, the inlet temperature and the fluid flow rate, resulting in nine scenarios tested with each panel. The experimental results obtained show that the heat dissipation ratios may vary between 106 and 227 W/m^2 in the case of the cooling panel and 140 and 413 W/m^2 for the solar collector. These results confirm the potential of the use of any of these panels to partially reduce the operation time of a cooling tower. From the empirical analysis, it must be addressed that the thermal conductivity of the bond between the pipes and the metal sheet is a key parameter to assure the highest heat dissipation possible in this serpentine type of panels. Regarding the model applicability, the average error is 4 % in the case of the cooling panel and 8.5 % in the case of the solar collector. Since the solar radiation of the model was calculated with monthly average values, the authors consider that this could be one of the parameters that most affect the error.

5. Acknowledgments

The authors would like to acknowledge the collaboration of Mikel Salazar from TROX SPAIN S.A., Junkers company (which is part of the Thermotechnology Division of Robert Bosch GmbH) and Juan Villaron from Tecnun (University of Navarra).

6. References

[1] European Commission 2016 *An EU Strategy on Heating and Cooling* (Brussels)

IOP Conf. Series: Earth and Environmental Science **154** (2018) 012008 doi:10.1088/1755-1315/154/1/012008

[2] M. Santamouris 2016 Cooling the buildings – past, present and future *Energy Build.* **128** 617–38

[3] M. Grosso, G. Fracastoro, M. Simonetti and G. Chiesa. 2015 A hybrid passive cooling wall system: Concept and laboratory testing results *Energy Procedia* **78** 79–84

[4] A. Zuazua-Ros, C. Martín-Gómez, J. C. Ramos and T. Gómez-Acebo. 2016 Bio-inspired heat dissipation system integrated in buildings: development and applications *Energy Procedia - 8th International Conference on Sustainability in Energy and Buildings* (Torino, Italy) pp 51–60

[5] Jamie Bartram, J. V. L. Yves Chartier and K. P. Surman-Lee and S. 2007 *Legionella and the prevention of legionellosis*

[6] ASHRAE 2003 Cooling Towers *ASHRAE Handbook* pp 1–68

[7] E. Erell and Y. Etzion 2000 Radiative cooling of buildings with flat-plate solar collectors *Build. Environ.* **35** 297–305

[8] M. Hanif, T. M. I. Mahlia, A. Zare, T. J. Saksahdan and H. S. C. Metselaar 2014 Potential energy savings by radiative cooling system for a building in tropical climate *Renew. Sustain. Energy Rev.* **32** 642–50

[9] Y. Cui, Y. Wang, Q. Huang and S. Wei 2016 Effect of radiation and convection heat transfer on cooling performance of radiative panel *Renew. Energy* **99** 10–7

[10] C. Yong, W. Yiping and Z. Li 2015 Performance analysis on a building-integrated solar heating and cooling panel *Renew. Energy* **74** 627–32

[11] T. Yamanashi, T. Hatori, Y. Ishihara, N. Kawashima and K. Niwa 2011 BIO SKIN Urban Cooling Facade *Archit. Des.* **81** 100–7

[12] A. Zuazua-Ros, C. Martín Gómez, J. C. Ramos and J. Bermejo-Busto 2017 Towards cooling systems integration in buildings: Experimental analysis of a heat dissipation panel *Renew. Sustain. Energy Rev.* **72** 73–82

[13] A. Zuazua-Ros, J. C. Ramos, C. Martín-Gómez and T. Gómez-Acebo 2017 Experimental assessment and model validation of a vertical cooling panel *Energy Build.* **142** 158–66

[14] J. A. Duffie and W. A. Beckman. 2013 *Solar engineering of thermal processes* (Wiley)

[15] Censolar 2007 *Distribución horaria de la irradiación solar global incidente sobre superficie horizontal en las cinco zonas climáticas definidas en el Código Técnico de la Edificación de España*

IOP Conf. Series: Earth and Environmental Science **154** (2018) 012009 doi:10.1088/1755-1315/154/1/012009

Passive designs and renewable energy systems optimization of a net zero energy building in Embrun/France

F Harkouss[1, 2], P H Biwole[3, 4] and F Fardoun[1]

[1] University Institute of Technology, Department GIM, Lebanese University, Saida, Lebanon
[2] Université Cote d'Azur, J.A. Dieudonné Laboratory, UMR CNRS 7351, Parc Valrose 06108 Nice, France
[3] Université Clermont Auvergne, CNRS, SIGMA Clermont, Institut Pascal, F-63000 Clermont-Ferrand, France
[4] MINES Paris Tech, PSL Research University, PERSEE - Center for Processes, Renewable Energies and Energy Systems, CS 10207, 06 904 Sophia Antipolis, France

Abstract. Buildings' optimization is a smart method to inspect the available design choices starting from passive strategies, to energy efficient systems and finally towards the adequate renewable energy system to be implemented. This paper outlines the methodology and the cost-effectiveness potential for optimizing the design of net-zero energy building in a French city; Embrun. The non-dominated sorting genetic algorithm is chosen in order to minimize thermal, electrical demands and life cycle cost while reaching the net zero energy balance; and thus getting the Pareto-front. Elimination and Choice Expressing the Reality decision making method is applied to the Pareto-front so as to obtain one optimal solution. A wide range of energy efficiency measures are investigated, besides solar energy systems are employed to produce required electricity and hot water for domestic purposes. The results indicate that the appropriate selection of the passive parameters is very important and critical in reducing the building energy consumption. The optimum design parameters yield to a decrease of building's thermal loads and life cycle cost by 32.96% and 14.47% respectively.

1. Introduction

Worldwide, buildings' energy demands are expected to increase in the following years. By the end of 2014, buildings (residential, commercial and public) consumed around 49% of the world's electricity consumption, where the residential sector consumed about 27% of the global electrical usage [1].

Net Zero Energy Building (NZEB) is a recent approach suggested to limit energy consumption and pollution emissions in buildings [2]. Building optimization is an effective method to evaluate design choices and to get the perfect solution for a specific purpose expressed as objective functions under several constraints [3]. Multi-objective optimization (MOO) results are sets of non-dominated solutions represented as a Pareto front [4]. After obtaining the Pareto front, the multi-criterion decision-making (MCDM) process is applied in order to select the final optimal solution among all offered possibilities. MCDM is capable to review the problem in accordance with the significance of different criteria and the preferences of the decision maker (DM).

In this paper, cost-effective design options for a residential NZEB in Embrun, a commune in the Hautes-Alpes department in the Provence-Alpes-Côte d'Azur region in South-eastern France, is inspected. Initially, the base case design options, RE systems, and simulation results are defined. Afterward, optimization problem of an extensive variety of design options is offered; including wall and roof insulation levels, windows glazing type, window to wall ratio (WWR) in eastern and western facades, cooling and heating set points, Photovoltaic (PV) and Solar domestic hot water system (SDHW) sizing. The optimization is carried out using "MOBO", a Multi-Objective Building Optimization tool presented by Palonen et al. (2013) [5]. Furthermore, in order to obtain a single solution, the ELECTRE III (Elimination and Choice Expressing the Reality) MCDM method, developed by Roy [6] is employed. Finally, a set of recommendations is outlined in order to ameliorate the performance design of NZEB in Embrun.

2. Base case building characterization

The investigated building is located in Embrun, characterized by an inland mountain climate. The monthly average dry bulb temperature is represented in 'Figure 1' according to the data from meteorological stations [7].

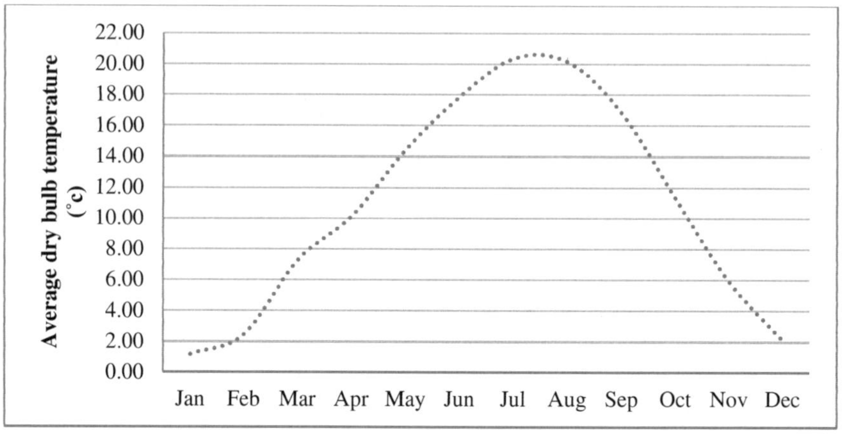

Figure 1. Monthly average dry bulb temperature for Embrun (Data Source: [7])

Besides, the building is composed of three stories, each of 205 m^2. Windows are distributed on Eastern and western facades, shaded with opaque blinds. Heating loads are covered by a natural gas condensing boiler (Efficiency=98.3%). Cooling loads are covered by air source heat pumps, with a coefficient of performance (COP) equal to 2.9. Systems set points are fixed respectively at 24°C for cooling and 20°C for heating during occupied hours. Though, during uninhabited times, cooling and heating systems are switched off. The design relative humidity is fixed at 50 %. The building is considered as tight, so the infiltration rate is equal to 0.38 ACH [8].

3. Renewable energy (RE) systems characteristics and simulation

3.1. Solar domestic hot water system characteristics

The hot water demands are covered by means of a flat plate direct active SDHW system with auxiliary electric heater. The SDHW system is composed of 5 solar collectors (SC) connected in series of total area equal to 31.35 m^2, south oriented, located at building rooftop and sloped to Embrun local latitude. The employed system characteristics are represented in 'Table 1'.

Table 1. SDHW system characteristics (Data source: [9])

Characteristics	Value	Characteristics	Value
Collector area, m^2	2.09	Collector flow rate, Kg/hr	70
Storage tank area, m^3	2.271	Intercept efficiency	0.79
Hot water set point, °C	60	Efficiency slope, $W/m^2.K$	3.48
Hot water supply temperature, °C	45	Efficiency curvature, W/m^2K^2	0.0056

3.2. Base case demands

The software TRNSYS is adopted to simulate buildings' electrical, thermal and hot water demands. The resulting buildings' total annual electrical demands are 64.03 KWh/y.m^2. Whereas, heating and cooling demands are 107.59 KWh/y.m^2 and 10.45 KWh/y.m^2 respectively. The monthly space heating and cooling demands are represented consecutively in 'Figure 2' and 'Figure 3'.

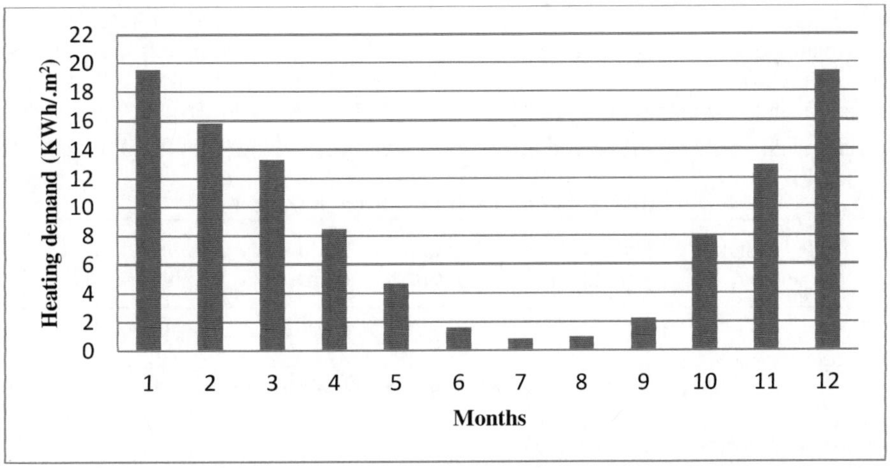

Figure 2. Monthly space heating load (KWh/m2 of heated area)

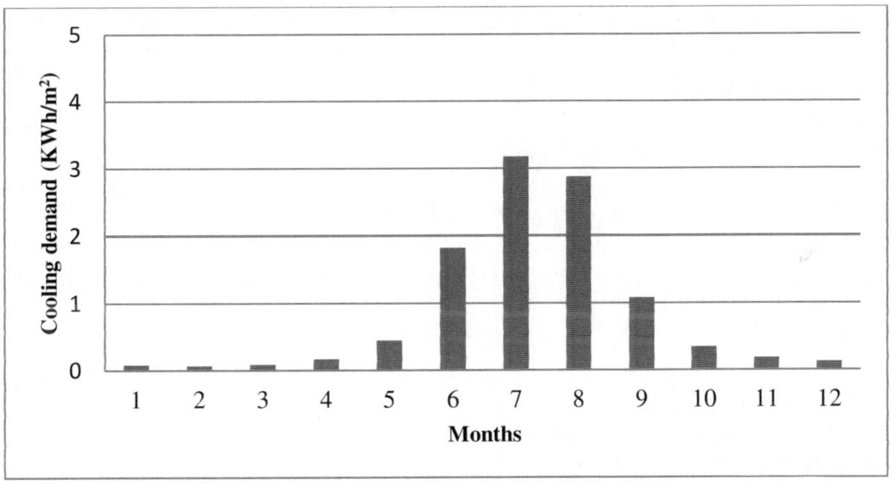

Figure 3. Monthly space cooling load (KWh/m2 of cooled area)

3.3. Photovoltaic system characteristics
In order to generate buildings required electricity, a PV system is used. The PV array is south oriented and sloped at Embrun local latitude. The analytical calculation [9], yield to 90 PV modules (15 in series, and 6 in parallel). The technical data of each module are represented in 'Table 2'. The building uses the utility power grid for storage, it delivers energy to the grid when the PV system produces more energy than the building needs and draws from the grid when the PV system produces less energy than the building needs.

3.4. Base case electrical and economic simulation results
The annual electric balances, are summarized in '

Table **3**'. Besides, the simulated LCC for a life period of 20 years and an annual discount rate of 5% is 169,300 \$ (275 \$/m2). It can be observed that about 38% of building loads are covered by the PV system. The zero energy balance is reached however with a great "Gains" that represents a loss in terms of investment cost. Also it is worth to mention the building's high heating load.

Table 2. Parameters of PV module (Data source:[10])

Panel characteristics	Value	Panel characteristics	Value
Short circuit current, A	9.32	Open circuit voltage, V	45.92
Current at maximum power, A	8.85	Number of cells in series	72
Voltage at maximum power, V	37.38	Panel area, m^2	1.94
Temperature coefficient/ open circuit voltage, V/K	-0.318	Module efficiency, %	17
Temperature coefficient/ short circuit current, A/K	0.042	Nominal output, Wp	295.3

Table 3. Summary of electrical balances

Description	PV system output	From PV to building	From PV to grid	Building demand	From grid to building	Gains "Load-generated by PV"
Value (MWh)	68.15	15.3	37.68	39.38	24.08	-27.14

Optimization problem and decision making process

3.5. Objective functions and decision variables

The electric consumption is due to thermal loads, lights, appliances and consumption of SDHW system auxiliary heater and circulating pump. The consumption from cooling and heating are minimalized when minimalizing the thermal demands. Besides, energy star appliances and fluorescent lights are used to reduce consumption. Therefore, in this study there are four objective functions to be minimized:

f1=min ("Auxiliary electric heater + Pump" consumptions)
f2=min (Thermal demand)
f3=min (Difference between load and generation)
f4=min (LCC)

'Table 4' provides the list of building passive and RE systems decision variables. The optimization problem is constrained by the occupant's thermal comfort, which is given by the following condition: the average predicted mean vote (PMV) of the whole building must remain between its acceptable bounds $|PMV| \leq 0.5$.

Table 4. Description and different options of decision variables used in the optimization problem

Description	Type	Values	Step
External walls, Roof insulation thickness (cm)	Discrete	1,3,5,7,10	-
Type of double glazing: Krypton or Argon, U-value ($W/m^2.K$)	Discrete	0.86, 1.4	-
Cooling set point (°C)	Discrete	24, 25, 26	-
Heating set point (°C)	Discrete	19, 20	-
Number of solar collectors in series	Continuous	1 to 20	1
SDHW pump flow rate (Kg/h)	Continuous	50 to 120	5
Number of solar panels in series	Continuous	1 to 20	1
Number of solar panels in parallel	Continuous	1 to 40	1
Width window bedrooms, kitchen (m)	Continuous	1 to 2	0.25
Width window Living and dining (m)	Continuous	1 to 3	0.25
Width window Living and dining (m)	Continuous	1 to 3.7	0.25

3.6. Building optimization tool, algorithm and Pareto front

In this study, the optimization tool MOBO is coupled with TRNSYS. The non-dominated sorting genetic algorithm (NSGA-II), developed by Deb et al. [11], is adopted. The parameters' of NSGA are set as follows: Population size = 40, Generation number = 25, Crossover probability= 70%, and Mutation probability= 2%. These parameters are selected based on the preliminary researches to get the best compromise between the Pareto-front accuracy and the optimization computational time [4]. Note that the four-objective optimization generates a four-dimensional (4D) problem. The projection of the 4D problem in a bi-dimensional (2D) graph, gives the Pareto front represented in 'Figure 4'.

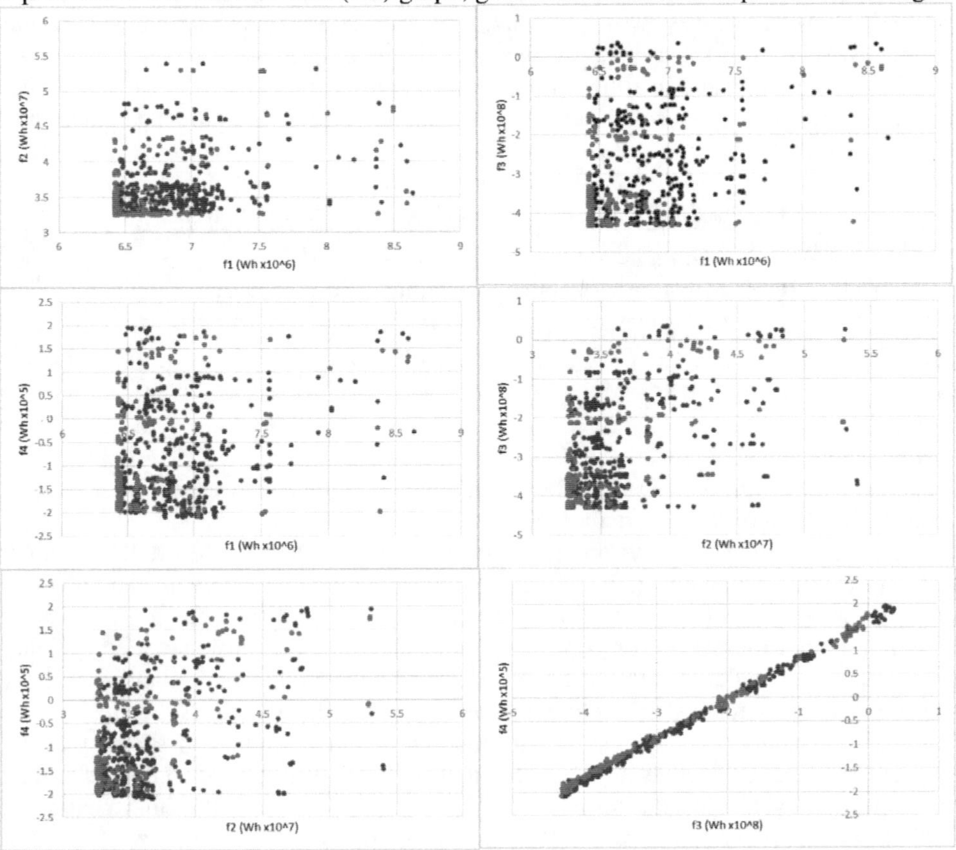

Figure 4. Bi-dimensional projections of the analyzed 4D-problem space (Blue: Building variants, Red: Pareto-front)

3.7. Decision making and sensitivity analysis

In order to classify the Pareto front solutions and to choose the most suitable one, the ELECTRE III method is adopted. Indifference, preference and veto thresholds of ELECTRE III are assigned respectively as follows: 5%, 10% and 30% relative to the average of each objective function's value for the Pareto front points. The Analytical Hierarchy Process (AHP), proposed by Saaty [12], is implemented to assign the relative weight of each objective. The best solutions resultant of ELECTRE III ranking are summarized in 'Table 5', together with the difference between the best case and the base case.

Table 5. Difference between best solution and the base case

	f1 (MWh)	**f2 (MWh)**	**f3 (MWh)**	**f4 (1000$)**
Best solution	6.73	34.12	-13.6	197.94
Base case value	6.98	50.90	-27.14	169.30
% difference	3.58	32.96	-49.92	14.47

3.8. Results and discussion

'**Table 6**' lists NZEBs parameters results of the decision making process. The results obviously show that there is a significant potential to improve the energy performance of residential buildings in Embrun, by implementing proven passive strategies. It can be noticed that building's envelop insulation is a key parameter, walls and roof insulation thicknesses are increased up to 7 cm and 10cm instead of 5 cm and 1cm respectively. Windows U-value of 0.86 W/m2.K is found to be more suitable than 1.4W/m2.K so as to lessen heat thermal flows from inside to outside the building. Moreover, the optimal WWR in eastern and western façades must ensure the essential heat gains from day time's solar radiation. WWR in eastern and western façades are decreased to 17.18% and 32.92% respectively. Optimal cooling and heating set points are modified to 25°C and 19°C respectively, with the intention of decreasing the thermal loads; the occupants comfort is untouched since the PMV is considered as a constraint.

Table 6. Summary of the optimal building design options

Variable	Insulation		Windows	Set points		WWR		PV	SDHW	
	Walls	Roof	U-value	Cooling	Heating	East	West		Pump flow	SC
Unit	(cm)	(cm)	(W/m^2.K)	(°C)	(°C)	(%)	(%)	-	(Kg/h)	-
Base case	5	1	1.4	24	20	23.43	59.5	108	70	15
Optimal	7	10	0.86	25	19	17.18	32.92	90	95	16

4. Conclusion

The challenge in ZEB design is to find the best combination of design strategies that would face the energy performance problems of a particular building.

In this study, an energy simulation and optimization programs (TRNSYS and MOBO) coupled with a ranking decision making technique (ELECTRE III) are employed to assess the most cost-effective passive strategies and RE systems sizes that should be implemented to achieve a NZE design for a residential building in Embrun. In the optimization analysis, an extensive variety of design and operating choices are considered including wall and roof insulation levels, windows glazing type, WWR in eastern and western facades, cooling and heating set points, PV and SC systems sizing.

The optimum design parameters and their corresponding objective functions shows that the annual thermal load and LCC can be decreased by 32.96% and 14.47% respectively, compared to the baseline model. Envelop high level of insulation is an essential step to decrease the high heating demands. Besides, WWR of about 17.18% and 32.92% respectively on eastern and western facades are sufficient to collect the necessary solar radiation and to avoid heat losses due to the excessive outside low temperature.

Future studies may be extended to other passive design parameters, energy efficient systems, and RE systems as well as other objective functions such as the life cycle assessment impact.

5. Acknowledgments

The authors would like to thank the Lebanese University, Cote d'Azur University, Campus France as well as the association of specialization and scientific orientation for their support.

6. References

[1] Anon 2016 International Energy Agency Statistics. http://www.iea.org.

[2] S. Pless and P. Torcellini 2010 *Net-Zero Energy Buildings A Classification System Based on Renewable Energy Supply Options* (National Renewable Energy Laboratory)

[3] A. Athienitis and W. O'Brien2015 *Modeling, Design, and Optimization of Net-Zero Energy Buildings* (Ernst & Sohn)

[4] N. Delgarm, B. Sajadi, S. Delgarm and F. Kowsary 2016. A novel approach for the simulation-based optimization of the buildings energy consumption using NSGA-II: Case study in Iran *Energy Build.*

[5] M. Palonen, M. Hamdy and A. Hasan 2013 Mobo a new software for multi-objective building performance optimization *Proceedings of the 13th International Conference of the IBPSA* pp 2567–2574

[6] B. Roy 1991 The outranking approach and the foundations of ELECTRE methods *Theory and Decision* **31(1)** 49–73

[7] Anon 2016 Meteo France

[8] ASHRAE 2001 ASHRAE Fundamentals Handbook (SI)

[9] Alamsyah, TMI, Sopian K and Shahrir A 2003 Techno-economics analysis of a photovoltaic system to provide electricity for a household in Malaysia. In: The international symposium on renewable energy : environment protection & energy solution for sustainable development. Kualalumpur, Malaysia 387–96

[10] Anon 2016 Mitsubishi electric, USA-solar innovations

[11] Deb K, Pratap A and Agarwal S 2002 A fast and elitist multi-objective genetic algorithm: NSGA-II *IEEE Transactions on Evolutionary Computation* **6** 182–97

[12] Saaty R 1987 The Analytical Hierarchy Process-What it is and how it is used *Mathl Modelling* **9** 161–76

IOP Conf. Series: Earth and Environmental Science **154** (2018) 012010 doi:10.1088/1755-1315/154/1/012010

A comparative study of fluorescent and LED lighting in industrial facilities

C Perdahci[1] PhD, H C Akin[2] BSc and O Cekic[2] MSc

[1] Electrical Engineering Department, University of Kocaeli, Kocaeli, Turkey
[2] Alkan Asia Africa Lighting Co., Fevzi Alkan Street, No:43 Sile, Turkey

Abstract. Industrial facilities have always been in search for reducing outgoings and minimizing energy consumption. Rapid developments in lighting technology require more energy efficient solutions not only for industries but also for many sectors and for households. Addition of solid-state technology has brought LED lamps into play and with LED lamp usage, efficacy level has reached its current values. Lighting systems which uses fluorescent and LED lamps have become the prior choice for many industrial facilities. This paper presents a comparative study about fluorescent and LED based indoor lighting systems for a warehouse building in an industrial facility in terms of lighting distribution values, colour rendering, power consumption, energy efficiency and visual comfort. Both scenarios have been modelled and simulated by using Relux and photometric data for the luminaires have been gathered by conducting tests and measurements in an accredited laboratory.

1. Introduction

There have been rapid and significant changes in lighting technologies since the invention of incandescent lamps. These changes and inventions mostly aimed to provide more efficient and higher quality lighting systems for different areas and usage [1]. Fluorescent lamps consume approximately %65 less energy when compared to incandescent lamps in the same luminous flux level. Energy efficiency has not been the only advantage of fluorescent lamps, along with this huge benefit, fluorescent lamps has also a considerably larger lifetime than regular incandescent lamps [2]. Fluorescent lamps also dissipate less heat, which is beneficial for the systems that are sensitive to heat. Fluorescent lamps contain vaporized mercury which is ionised with applied voltage across the tube. Ionisation of the mercury vapour requires a higher voltage than the mains voltage; therefore, a starter and a ballast are necessary to provide the required voltage for initial ionisation. After ionisation occurs and fluorescent lamb starts operating, the lamp shows negative resistance effect. The conductivity of the lamb increases with the current flows through it. In order to solve this issue, ballast must be used in order to limit the current by adding inductive resistance into the circuit [3].

Fluorescent lamps require ballasts to operate which causes additional power loss and also increases the initial cost [3]. Ballasts can be magnetic, electronic and hybrid. Magnetic ballasts contain magnetic core wrapped copper windings to regulate the current and voltage based on the principle of magnetic induction [4]. Magnetic ballasts work with line frequency which causes a higher flicker index. On the other hand, since they don't contain semi-conductors, harmonic emission is not an issue for magnetic ballasts when comparing to electronic ballasts. Electronic ballasts work in a different way. Advanced technology in electronics allowed to use semi-conductors in ballasts. Thus, magnetic core and windings were replaced with electronic components. With the addition of semi-conductors, frequency control became possible for electronic ballasts [5]. A comparison of magnetic and electronic ballasts can be found in Table 1.

Table 1. Comparison between magnetic and electronic ballasts

Advantages of Magnetic Ballasts	Advantages of Electronic Ballasts
Considerably lesser harmonic emission	High energy efficiency
Able to operate in tough conditions	High power factor
Failures/malfunctions occur less since magnetic ballasts do not have delicate components	Extend lifetime of the lamp %20-25
Lesser manufacturing costs	Compatible with dimmable systems
EMC is not an issue since the operating frequency is low (mains frequency)	Compatible with automation applications

LED lamps have been developed and became popular rapidly with the evolution of technology. When LED lamps appeared in the market, they were inefficient; however, developments in LED technology have taken the LED lamps to top of the energy efficient lighting solutions [6]. In order to use LED lamps with mains voltage and current or the DC voltage and current as needed, LED drivers are required. LED drivers are basically AC/DC or DC/DC converters. Drivers contain semi-conductors which cause massive harmonic emissions; therefore, LED drivers are being used with PFC [7].

2. Illuminating industrial facilities

Lighting quality is one of the most essential elements in a work place. Complex and large industrial facilities have different lighting requirements for different parts of the facility [8]. Lighting requirements of a facility also changes with personal tasks. These requirements have been set by certain standards. EN 12464-1 standard sets the minimum lighting needs in a work place.

In industrial facilities, increasing energy efficiency has always been the top priority to maximize profit and to help preserving natural resources [9]. With the scientific breakthroughs, advancing technology and new inventions, many ways of increasing energy efficiency have been discovered and developed. Automation technology, using new methods and materials for isolation, energy efficient motors and energy efficient lamps are the most effective ways to decrease energy consumption [9], [10].

Among all methods which aim to increase energy efficiency, switching into a more efficient lighting system is mostly the cheapest and the most effortless method. Depending on the luminaire type, quantity and price, initial costs can be different. Installation costs, maintenance costs and labor costs during installation must be considered as initial cost [10].

Lighting requirements for buildings used as warehouses depend on the usage of the building and personal tasks. A warehouse which contain forklift works, needs a clear vision for safety and performance reasons. Moreover, warehouses with packaging and loading jobs need a higher lux value. As can be seen in table 2, lighting standards for a warehouse can change depending on the task and the usage of the building according to EN 12464-1.

The photometric tests and measurements of the luminaires have been conducted in an accredited laboratory and the necessary luminaire data have been acquired. Both scenarios have been built based on a warehouse in İstanbul by considering require lighting standards and visual comfort. Figure 1. shows the luminaires which have been used in the study. In order to perform simulation, the warehouse have been modelled in Relux software.

3. Case study

The case study has been conducted in a warehouse of a factory in Istanbul. The dimensions of the warehouse are 20 m for length, 10 m for width and 8 m for height respectively. The height of the work plane is 0.75 m and the distance between floor and the luminaire is 7.5 m.

The reflection coefficient rates of the walls are ρ wall=0.5, floor ρ floor=0.2, and the ρ ceiling=0.7.

As seen on the Table 2, the required lux level which can provide minimum illumination needs for the warehouse mentioned must be above 300 lx.

The maintenance factor has been assumed as 0.8.

Figure 1. 53. AKYT 254 (fluorescent) and 53.AKYTL 220 (LED) luminaires

Table 2. EN 12464-1 lighting standards for a warehouse

Interiors Activities	Em[lx]	Ra	UGRL
Store room aisles and passages	100	60	25
Warehouses with uniform or bulky items	300	60	25
Warehouses with various items and search requirements	300	60	25
Warehouses with reading requirements	300	60	25

In this study, the economical and efficient ways have been researched in two different scenarios by using two different lighting solutions in order to provide minimum light requirements for visual comfort and visual task according to ISO 8995-1:2002(E)/CIE S 008/E:2001 standard. In scenario 1, the storage room has been designed by using 12 pieces of Alkan 53.AKYT 254 luminaire with TL 5 fluorescent lamp. The photometric data of the luminaire which has been used in scenario 1 according to test results have been shown in Figure 1. In scenario 2, the warehouse has been designed by using 12 pieces of Alkan 53.AKYTL 220 LED luminaire. The photometric data that have been acquired from test results of the luminaire which has been used in scenario 2 have been shown in Figure 6. The facility where experiment has been conducted operates 7 days and 24 hours.

The lighting scheme for the study has been designed by using Relux simulation software and the simulation results are presented for each scenario.

3.1 Scenario 1

An Alkan product 53.AKYT 254, which contains 2X54W fluorescent lamps, is used for the illumination of the warehouse in the facility. The luminous flux of used luminaire is 7120 lumen (lm) and the power is 113,5 W. The luminous flux of each lamp in luminaire is 4450 lumen. The efficiency of the luminaire can be calculated as;

$$\frac{7120\ \mathrm{lm}}{4450\ \mathrm{lm} \times 2} = \%80$$

Totally 12 luminaires were used for the study. The total luminous flux is the sum of all the flux emitted by 12 luminaires and equals 85440 lm, also the total used power equals to;

$$Pt = n.P = 12x113{,}5W = 1362W$$

According to result of Relux program, the average illuminance measured at work plane Eav=232 lux, the minimum illuminance measured at work plane Emin=156 lux and the maximum illuminance measured at work plane Emax=277 lux.

Figure 2 presents photometric data for scenario 1. The positioning of the luminaires can be seen in Figure 3. The lighting distribution values in Lux are illustrated in Figure 4. Figure 5 is 3-D colour rendering and presents the effect of lighting on different areas in the room. Figure 6 presents false colour rendering scheme.

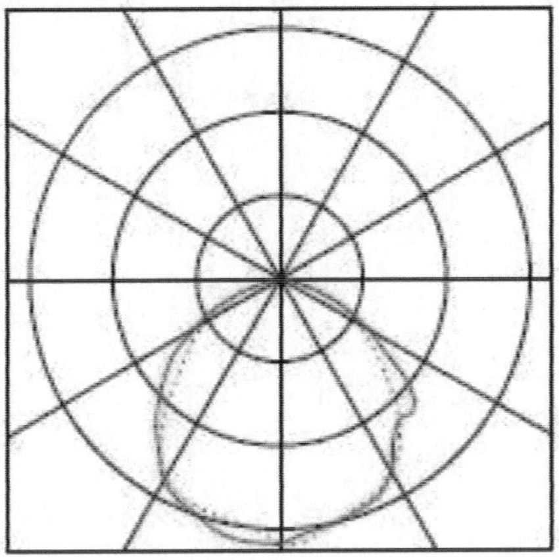

Luminaire efficiency	80%	**Luminaire ID**	53.AKYT 254
Luminaire efficacy	69.13 lm/W	**Luminous flux of the lamps**	4450 lm
CIE Flux Codes	53 85 98 100 80	**Power of the lamps**	54 W
UGR 4H 8H	22.3 / 21.2	**Whole circuit power**	113,5 W
Luminous flux of the luminaire	7120 lm	**Quantity of the lamps**	2

Figure 2. Polar diagram and photometric data of the luminaire 53.AKYT 254

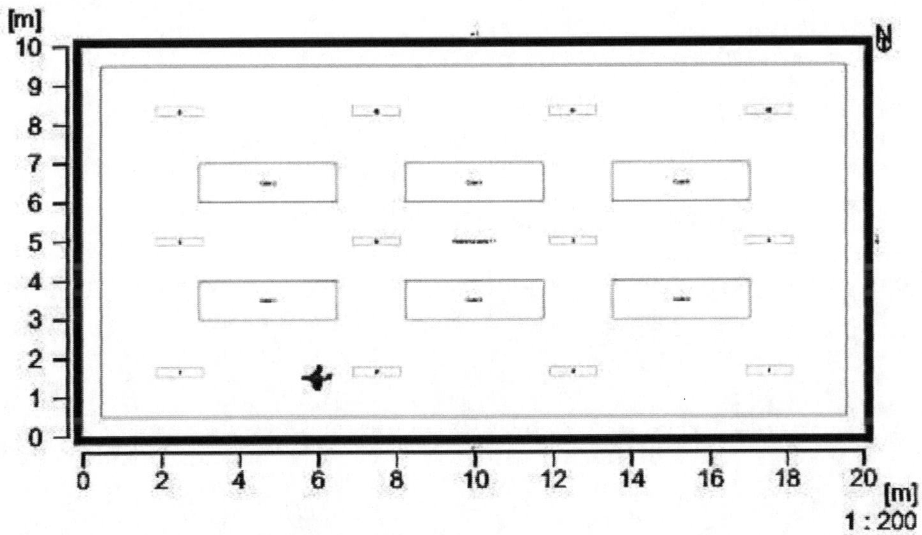

Figure 3. Layout plan for Scenario 1

3.2 Scenario 2

Alkan product 53.AKYTL 220 LED Panel is used for the illumination of the warehouse for the concerned facility.

The luminous flux of each LED luminaire is 11450 lm and the power is 112W. The luminous flux of the LED lamps in each luminaire is 14800 lm. The efficiency of the luminaire can be calculated as;

$$\frac{11450 \text{ lm}}{14800 \text{ lm}} = \%77$$

Totally 12 luminaires were used for the study. The total luminous flux is the sum of all the flux emitted by 12 luminaires and equals 137400 lumen. Also the total used power equals;

$$Pt = n.P = 12x112\ W = 1344\ W$$

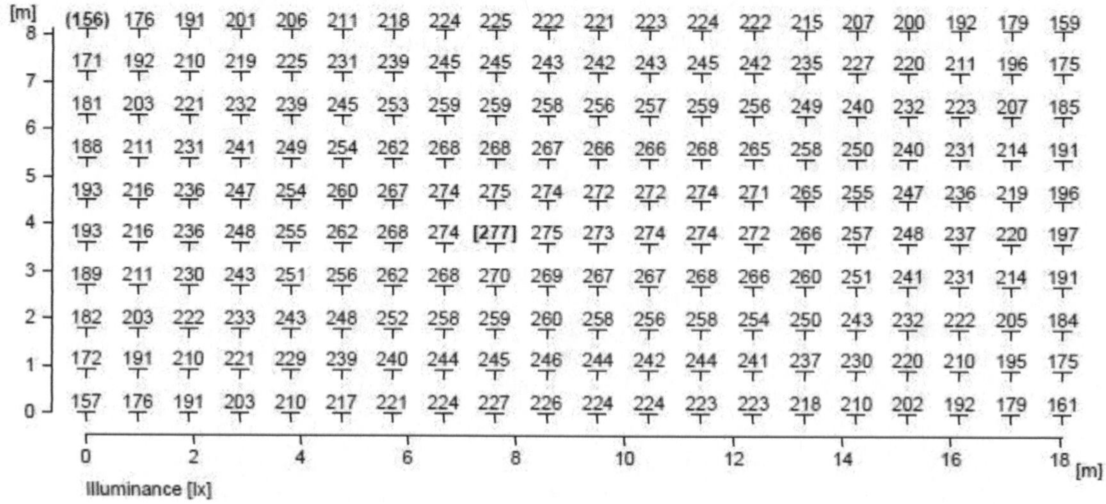

Evaluation area 1	Reference plane 1.1	
	Horizontal	
Em	232 lx	
Emin	156 lx	
Emin/Eav (Uo)	0.67	
Emin/Emax (Ud)	0.56	
UGR (1.6H 3.2H)	<=21.3	
Position	0.75 m	
Major surfaces	**Em**	**Uo**
m 1.5 (Ceiling)	55 lx	0.83
m 1.1 (Wall)	143 lx	0.36
m 1.2 (Wall)	131 lx	0.41
m 1.3 (Wall)	140 lx	0.36
m 1.4 (Wall)	127 lx	0.40

Figure 4. Light distribution values for Scenario 1

Figure 5. Colour rendering for Scenario 1

According to result of Relux program, the average illuminance measured at work plane Eav=422 lux, the minimum illuminance measured at work plane Emin=270 lux and the maximum illuminance measured at work plane Emax=520 lux.

IOP Conf. Series: Earth and Environmental Science **154** (2018) 012010 doi:10.1088/1755-1315/154/1/012010

Figure 7 presents photometric data for scenario 2. The positioning of the luminaires can be seen in Figure 8. The lighting distribution values in Lux are illustrated in Figure 9. Figure 10 is 3-D colour rendering and presents the effect of lighting on different area in the room. Figure 11 presents false colour rendering scheme.

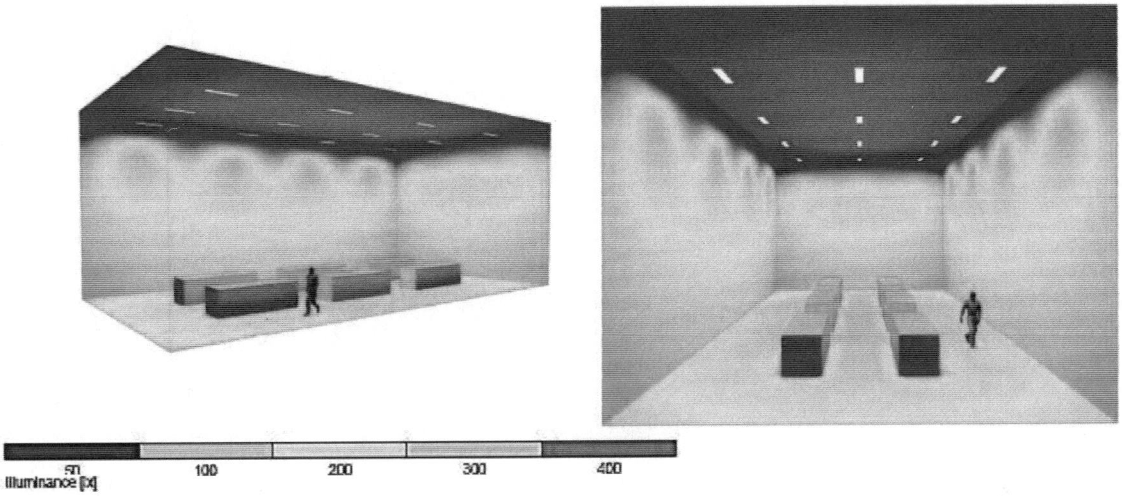

Figure 6. False colour rendering

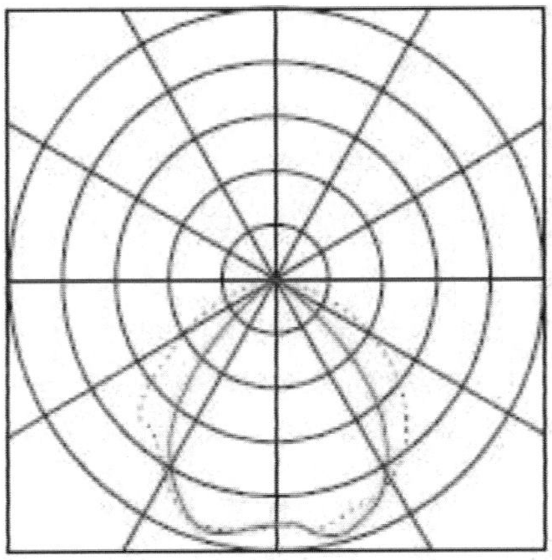

Luminaire efficiency	77%	**Luminaire ID**	53.AKYTL 220
Luminaire efficacy	102.23 lm/W	**Luminous flux of the lamps**	14800 lm
CIE Flux Codes	53 85 98 100 80	**Power of the lamps**	100 W
UGR 4H 8H	17.8 / 24.1	**Whole circuit power**	112 W
Luminous flux of the luminaire	11450 lm	**Quantity of the lamps**	240

Figure 7. Polar diagram and photometric data of the luminaire 53.AKYTL 220

4. Results and discussion

In this study, the lighting system of a workplace has been designed and analysed with 2 different approaches. The aim of both methods was to determine the energy efficient ways in lighting solutions. Fluorescent lamps are still being preferred in a lot of facilities because of the advantages such as efficiency and accessibility in open market. However, with the development in power electronics and technology, LED lamps are becoming more useful and efficient each day.

In designing process of the warehouse, fluorescent and LED luminaires have been placed on the same spots in order to collect consistent data and to make comparison more conclusive.

Illumination values for both scenarios have been shown in Table 2. It can be seen that LED lamps has higher efficacy than fluorescent lamps which shows that, when used in the same wattage, LED lamps have higher luminous flux than fluorescent lamps.

IOP Conf. Series: Earth and Environmental Science **154** (2018) 012010 doi:10.1088/1755-1315/154/1/012010

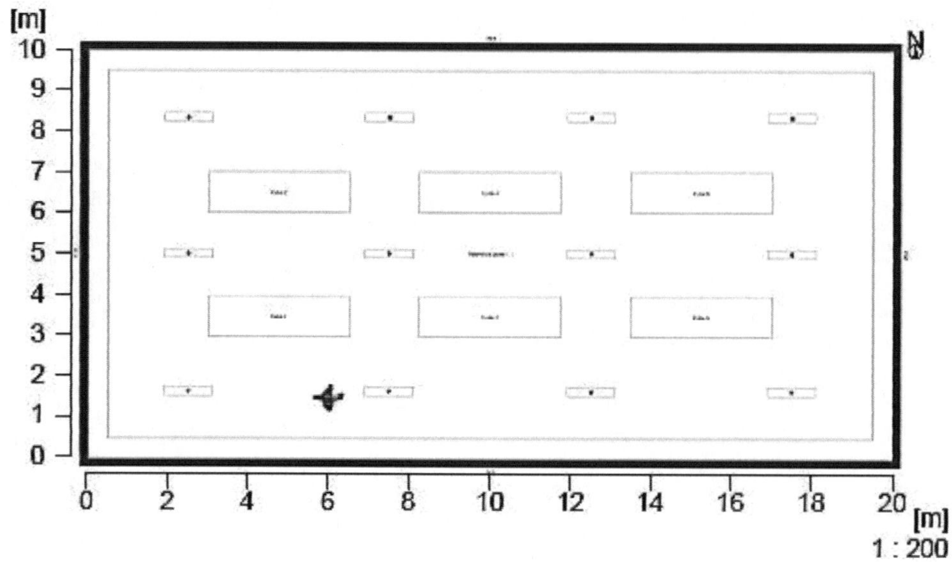

Figure 8. Layout plan

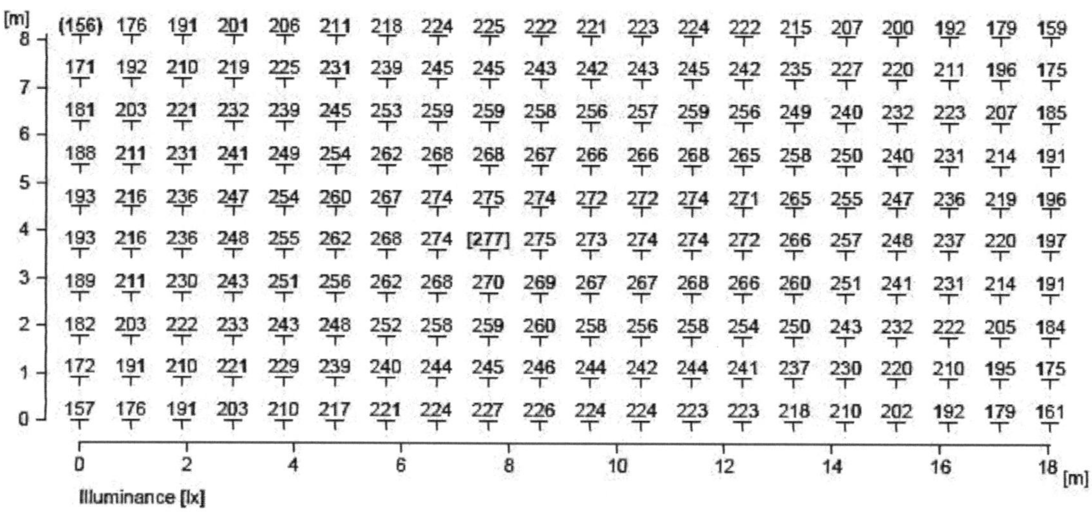

Figure 9. Lighting distribution for LED

Power consumption and emitted luminous flux can be seen together in Table 3. Luminous efficacy value can be reached by using luminous flux and power consumption data of a luminaire. In scenario 1, luminous flux has been calculated as;

$$7120\ \mathrm{lm}/113{,}5\ \mathrm{W} = 69{,}13\ \mathrm{lm/W}$$

As for scenario 2, luminous flux of the luminaire used is;

$$11450\mathrm{lm}/112\ \mathrm{W} = 102{,}23\ \mathrm{lm/W}$$

By referring to this data, it can be said that LED lamps can emit more luminous flux than fluorescent lamps in the same power consumption level because of the difference between their luminous efficacy values. When the same number of luminaires has been used, it has been seen that the system using fluorescent lamps cannot provide the required luminous flux for a healthy working environment according to international standards.

Figure 10. 3D Colour Render

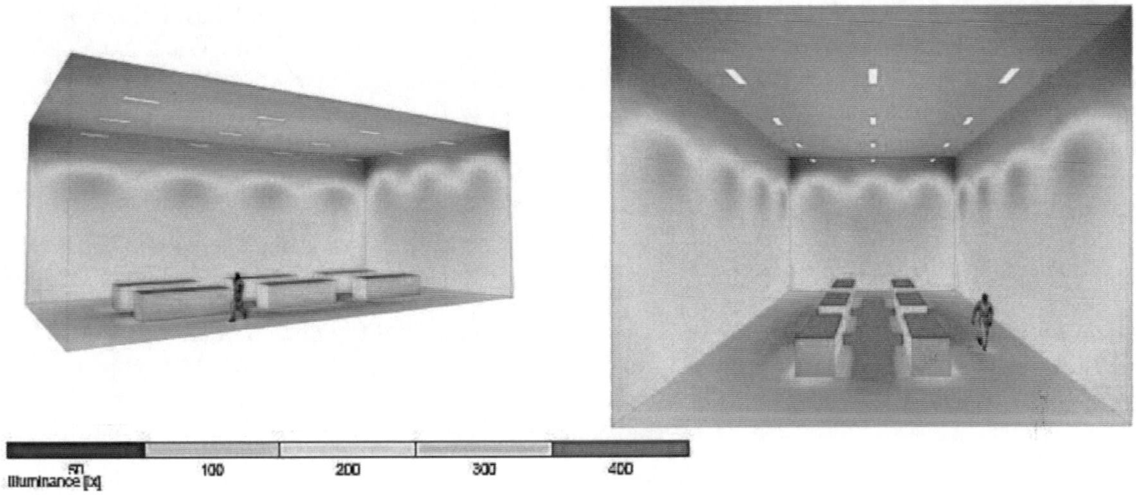

Figure 11. False Colour Render

Table 3. Illumination values for two scenarios

Type	Luminous Flux of Lamps	Luminous Flux of Luminaires	Luminous Efficacy	Total Luminous Flux (For 12 Luminaires)
Fluorescent	9900 lm	7120 lm	69,13 lm/W	85440 lm
LED	14800 lm	11450 lm	102,23 lm/W	137400 lm

Table 4. Power consumption and illumination values for two scenarios

Type	Power Consumed	Total Power Consumed	Luminaire Efficacy	Total Luminous Flux (For 12 Luminaires)
Fluorescent	113,5 W	1362 W	69,13 lm/W	85440 lm
LED	112 W	1344 W	102,23 lm/W	137400 lm

In order to reach the minimum lux level which can provide a healthy work place in scenario 1, more luminaires must be used and this design will have undesired consequences. Using more luminaires increases power consumption of the facility. In order to use more luminaires, more space will be required in warehouses, and placing luminaires can be problem since the room might be used at full capacity with products or wares that have been placed in the warehouse.

5. Acknowledgements

The authors would like to also thank University of Kocaeli and Alkan Asia Africa Lighting Co. for their valuable contributions to the present study.

6. References

[1] J. Schleich, B. Mills and E. Dütschke 2014 A Brighter Future? Quantifying the Rebound Effect in Energy Efficient Lighting *Elsevier* 72, pp.35-42

[2] D. Zhu and C. J. Humphreys 2016 Solid-State Lighting Based on Light Emitting Diode Technology *Optics in Our Time* chapter 5 p 112

[3] D. M. Babu, S. Regatti and S. G. Rao. 2013 An Electronic Ballast with U P F *International Journal of Latest Technology in Engineering, Management & Applied Science* Volume II, Issue IX pp 39-48

[4] W. M. Ng, D. Y. Lin and S. Y. Hui 2012 Design of a Single Ultra-Low-Loss Magnetic Ballast for a Wide Range of T5 High-Efficiency Fluorescent Lamps *Ieee Transactions On Industrial Electronics*, Vol. 59, No. 4, APRIL 2012

[5] I. Aliskan and R. Keskin 2016 Fluorescent Lamp Modelling and Electronic Ballast Design by the Support of Root Placement *International Journal of Engineering Technologies* Vol.2, No.3, 2016

[6] M. Cole, H. Clayton and K. Martin 2014 Solid State Lighting: The New Normal In Lighting Petroleum *and Chemical Industry Conference* (PCIC Brasil) pp 194-202

[7] R. K. Pandey and Asst. Prof. S. P. Tondare 2016 An AC LED Driver with Improved Total Harmonic Distortion and Power Factor *International Journal of Scientific & Engineering Research, Volume 7, Issue 5,* May-2016

[8] A. R. Musa, N. A. G. Abdullah, A. I. Che-Ania, N. M. Tawila and M. M. Tahira 2011 Procedia - Social and Behavioral *Sciences UKM Teaching and Learning Congress 2011*, vol. 60 pp. 318 – 324

[9] V. Bogatishchev. 2014 Improving the Effectiveness of Industrial Lighting *Metallurgist* Vol 58

[10] J. Lillo, H. Moreira, L. Álvaro and D. Majarín. 2013 Lighting and Work: Light for Seeing and Healthy Light *Psyecology*, 4:1 pp 11-38

IOP Conf. Series: Earth and Environmental Science **154** (2018) 012011 doi:10.1088/1755-1315/154/1/012011

Illumination of the marmaray project

C Perdahci[1], D Yuce[2] and R. A. Aynaci[3]

[1] University of Kocaeli, Kocaeli, Turkey
[2] Litpa Lighting, Istanbul, Turkey
[3] Litpa Lighting, Istanbul, Turkey

Abstract. The main objective of the underground railway stations lighting is to provide safe and comfortable conditions. This paper focuses on the illumination of Marmaray Project in order to shed light on long underground tunnel lighting design requirements and special railway tunnel environment. Moreover lighting of each Marmaray station and ticket hall were presented.

1. Introduction

With the increase in population (300,000 people) per year, Istanbul is the largest city in the Republic of Turkey. As the demand to the modern transportation systems increases in great cities like Istanbul, the fast growing and deepening problems in its transport infrastructure raise up. Because of the severe traffic in the morning and evening, the rush hours strings out which diminishes the options to increase the general capacity of the road systems. [1]

Istanbul is expanding its railway network to fulfill the needs of transportation. Therefore, The Marmaray Project can be considered as the corner stone of the transport system in Istanbul. There is going to be numerous transport facilities of the municipalities which will ease the transportation in the city. [1]

In the construction of The Marmaray Project, there are three bored tunnels and an immersed tunnel under the Istanbul Strait. The immersed tunnel is 1,4 km long, the connections between the immersed and adjacent tunnels included.

The most important lighting design factors for terminal and station lighting in public transport are: the appearance of space and lighting devices, the glare that the light from the source or reflection creates, security, field of view, horizontal and vertical lighting, integrated lighting and daylight control. [2]

Illumination of underground subway stations should allow safe circulation, provide user comfort, take into consideration the safety of the disabled, provide smooth transition from darker areas to bright areas. [2]

2. The marmaray project

Connecting the two banks of the Bosphorus has been a long term dream since 1502. It is rumored that the well-known artist Leonardo Da Vinci visited Istanbul as he had learned that Sultân Bayezid II had the idea for a bridge crossing the strait. The concept of a railway tunnel under the Bosphorus was first expressed by Sultan Abdülmecid in 1860s. American engineers studied the project in 1902 during the reign of Abdul Hamid and obtained a patent for Tünel-i Bahri.

A railway tunnel underneath the Bosphorus Strait has been opened in Turkey, connecting the Asian and European shores of Istanbul under the Marmara Sea. The Marmaray tunnel is the first tunnel connecting two continents in the world. The long desired dream to connect the two shores of Istanbul via an underwater tunnel came true with the help of this project.

The Marmaray Rail Tube Tunnel and Commuter Rail Mass Transit System project has started in the year 2004 in Istanbul.
A seamless high-capacity commuter rail system, an inter-city passenger service and a freight link between Europe and Asia were considered to be of utmost important characteristics of The Marmaray Project. The Gebze– Halkali section which is also a part of the Ankara–Istanbul corridor will meet the transportion needs of the city of Istanbul as well as its hinterland.
The line goes underground at Yedikule, continues through Yenikapı and Sirkeci via new underground stations, passes under the Bosphorus, connects to the new Üsküdar underground station and emerges at the surface at Söğütlüçeşme. [3] The entire system will be 76 km long, 13.4 km of which will be underground.
The other constructions of The Marmaray Project will be three underground stations, as well as 37 surface stations (36 of which will be new), an operational control center, yards, workshops and maintenance facilities.
There will be 37 stations in the Marmaray project. Only one of these will be reused, and the remaining 36 stations will be new constructions. To put it in a different way, 18 listed historic stations (Sirkeci, Kumkapı,Yenikapı, Kocamustafapaşa,Yedikule, Bakırköy, Yeşilköy, Haydarpaşa, Kızıltoprak, Feneryolu, Göztepe, Erenköy, Suadiye, Bostancı, Maltepe, Kartal, Pendik and Gebze), will be abandoned and lose their original function. [4]
The major objectives of the project are to:
a) find a long-term solutions to the present urban transport problems of Istanbul,
b) ease susbtantial operational problems on mainline railway services,
c) provide direct railway ties between Asia and Europe
d) increase capacity, reliability, accessibility, punctuality and safety on the commuter rail services,
e) decrease journey time and increase comfort for majority of commuter rail passangers,
f) provide uninterrupted passenger and freight transportation across the Istanbul Strait,
g) decrase air pollution which is caused by exhaust gasses, which will improve the air quality of Istanbul as a consequence,
h) reduce noise pollution resulting from airborne traffic in the centre of Istanbul, and
i) decrease the negative effects of car transport on historical buildings and heritage sites by providing a new alternative in the historical centre of Istanbul. [1]

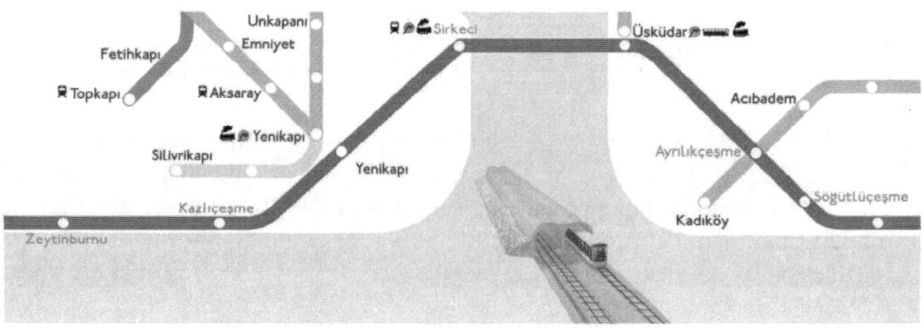

Figure 1:The Marmaray Project Stations

Because underground subway stations create different spaces within the urban fiction, artistic works are also included in the stationed areas in order to give identity to the stations or to establish relations with the city. [2] The Iznik Foundation Tiles have been used in this magnificent underground metro station that connects Asia and Europe.

3. Illumination

In underground subway stations; physical injuries such as injury to people, respiration, hearing and difficulty in seeing, as well as psychological stress, fear, anxiety and other negative effects are likely to occur. For this reason, it is important to provide visual comfort in underground metro stations. [2]

IOP Conf. Series: Earth and Environmental Science **154** (2018) 012011 doi:10.1088/1755-1315/154/1/012011

In such structures, the psychological pressure and fear experienced by the passengers can be reduced by the combination of light and color. A refreshing effect was observed at stations where the cool white or warm white light with a color temperature of 500 Kelvin was used for general illumination. In addition, red, yellow colors may be preferred in regional lighting to enhance attention. [2]

Figure 2: Üsküdar Station

Figure 3: Üsküdar Station

Reducing energy consumption while increasing profitability and improving visibility as well as safety were the main goals of the project. All luminaires should meet the photometric and electrical performance requirements and also they should be aesthetically appropriate to their intended location. These considerations are all taken into consideration while designing the luminaires.

Indirect lighting should be preferred in underground stations. It is possible to cause eye irritation if proper lighting is not attained. Eye dazzle, contrast can be avoided and comfortable vision can be achieved with proper lighting. From this point of view, the eye adaptation from the dark places to the bright places or from the bright places to the dark places can be obtained if the light transitions between different locations are not sharp.

3.1. Tunnels

Tunnel lighting should maintain required lighting level at the evacuation and maintenance walkway so that passengers can see the direction of escape and recognize the risk of tripping along the walkway edge. [5]

The main purposes of tunnel lighting can be listed as:

- Dropping off passengers safely and helping them walk along the tunnel.
- Helping maintenance personnel perform routine checks throughout the tunnel
- Guide train personnel and improve road visibility [6]

The system design where tunnel lighting is installed should provide the following:

- A luminous level of at least 10 Lux by considering the maintenance factor on the horizontal level on the rail level and a level of 2 Lux on the vertical level if it is 1 m above the rail level
- The accountability of the signaling
- A high level of lighting uniformity that facilitates adaptation
- A precise smoke, fire alarm system in the tunnel for emergencies.
- A spare supply source for safety in the event of a failure of the normal supply source.
- A minimum of 10 Lux horizontal lighting on the emergency walkways along the route even when the maintenance factor is added. [6]
- The general illumination of the Marmaray tunnel is performed using LITPA FLP-TV waterproof luminaires with protection class IP66. Burning point of the luminaires, which are made of polycarbonate V2 rated material, is higher than 850°C. This material is shatterproof and can resist impacts up to 5J.

IOP Conf. Series: Earth and Environmental Science **154** (2018) 012011 doi:10.1088/1755-1315/154/1/012011

Table 1: Lighting Criteria for Station Areas

Lighting Criteria for Station Areas According to National and International Standards					
	TS 12127 and DLH Criteria	GI/RT 7010 Railway Group Standard	EN 12464_1		
	Em	Em >	Min. Em	Max. UGR	Min. Ra
Traffic Zones					
Circulation Areas and Corridors	175-250		100	28	40
Stairs, Escalators, Moving Tapes	150-200		150	25	40
Railway Facilities					
Closed Platforms and Passage Underpasses	150-200		50	28	40
Ticket Hall and Conkors Floor	200-400		200	28	40
Ticket And Luggage Offices And Turnstiles			300	19	80
Waiting Rooms			200	22	80
Platform Edge					
Platform Edge Zone, horizontal lighting	250-300	10			
Platform Edge Area, vertical luminous level		2			
Tunnel Lighting					
Rail Upper Thick, horizontal plane		10			
Vertical plane 1m above Rail Top Coat		2			
Authorized Walkway		10			
Emergency exit and escape routes, vertical lighting		2			
Emergency exit and escape routes, horizontal lighting		10			

A critical area to be lit along a metro route is the so-called yellow line that is used to regulate boarding and disembarking the trains. The international standard EN12464-1/8.2.1 requires that such places should have a minimum illumination level of 150 lux. Currently the illumination level is 400 lux. Taking into consideration the maintenance factors of long term operation, a minimum of 340 lux should be ensured.

The luminaires used along the tunnels of the Marmaray are resistant to salt and moisture of the sea as well as vibration and pressure effects. The luminaires in the tunnels are also designed to be resistant to 5J impacts. In order to prevent combustion of luminaires and the spread of fire, V2 polycarbonate material was used to manufacture the fixtures and the flash-point temperature was raised to 850°C.

236 FLP-TV fixtures are placed every 6 meters along the tunnel to obtain an illumination level of 50 lux throughout.

Figure 4: The Marmaray Tunnel with LITPA FLP-TV Luminaires

Figure 5: Luminaires along Yellow Lines

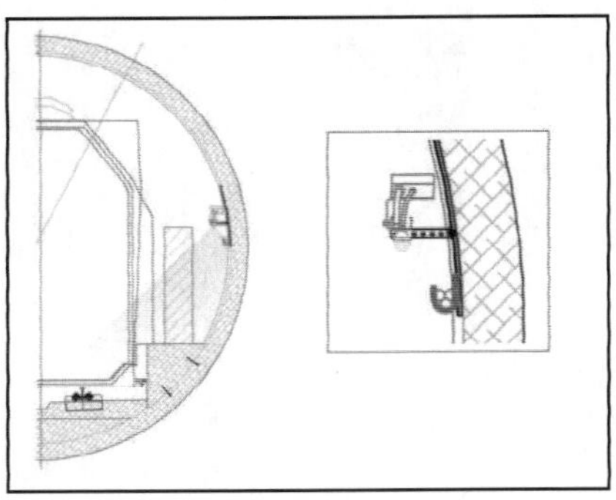

Figure 6: Luminaire Layout in the Tunnel

3.2. Ticket Halls

Ticket hall lighting should be bright and welcoming. Moreover, the key customer touch points such as; ticketing, information areas and gate lines should be highlighted. Orientation lighting to draw customers toward route ways for their onward journey is also essential.

Applying ceiling recessed luminaires has helped to wash the key vertical surfaces of the space, define the space and make it feel bright and welcoming.

The ticket halls and transit points have an illumination level of 310 lux, well in excess of the minimal level required by the international standard EN12464-1/8.2.2 which is 200 lux. For luminaires with protection class IP65, resistance to ambient conditions is increased so as to minimize potential lighting problems due to contamination of fittings over time. An illumination level of up to 400 lux is achieved near the stairs where pedestrian traffic is very heavy.

Figure 7: Ticket Halls

3.3. Stations

Lighting of each Marmaray station was handled separately. The types of luminaires used in each of the stations and tunnels were determined case by case.

Figure 8: Stations

The Marmaray includes five stations: Kazlıçeşme, Yenikapı, Sirkeci, Üsküdar and Ayrılıkçeşme. The luminaires used for lighting purposes were manufactured to resist all unfavorable conditions.

They have been designed designed to go beyond the international standards for ambient lighting, light quality, and fixture quality levels.

Figure 9: The LITPA Luminaires in the Marmaray Project

4. Conclusion

In underground railway stations, lighting is one of the most significant properties to achieve and sustain safety. Compatible with the new emphasis in environmental concerns, a design approach which is highly energy saving and cost-effective is essential.

In the study, The Marmaray Project was selected as the sample underground railway station. First of all, The Marmaray Project is introduced and then lighting criteria for underground railway station areas according to national and international Standards are presented. Additionally, tunnel lighting and lighting of each Marmaray station and ticket hall were explained.

5. Acknowledgments

The authors would like to thank University of Kocaeli and LITPA Lighting Co. for their valuable contributions to the present study.

6. References

[1] H. Belkaya, Dr. I. H. Ozmen, Dr. I.. Karamut. The Marmaray Project: Managing a Large Scale Project with Various Stake Holders, *Proceedings of the World Congress on Engineering 2008* Vol II WCE 2008, July 2 - 4, 2008.

[2] D. Aktop Maden, E. Avlar, *Evaluation of the Passenger Areas of Underground Subway Stations in Terms of Visual Comfort: The Sample of Kadiköy and Kartal Stations*, MEGARON 2017;12(1):13-26.

[3] S. Lykke, and H. Belkaya. 2005. Marmaray Project: The Project and its management. *Tunelling and Underground Space Technology 2006*: 600–603.

[4] Y. Erkan, (2012). The Railway Heritage of Istanbul and the Marmaray Project", *International Journal of Architectural Heritage*, 6(1), 2012, 86-99.

[5] H. M. S. Leung, LED Lighting Application in Long Railway Tunnels" ,The Hong Kong Institution of Engineers Transactions, vol.19, No.4, pp 42-46, 2012.

[6] S. Tuna. Yeraltı Metro Sistemlerinin Aydınlatılmasında Enerji Etkinliği, January 2008.

IOP Conf. Series: Earth and Environmental Science **154** (2018) 012012 doi:10.1088/1755-1315/154/1/012012

Viability of using energy storage for frequency regulation on power grid

Y S Lim[1,3], L C Hau[1], K Y Loh[1], K Y Lim[1], P F Lyons[2] and P C Taylor[2]

[1]Universiti Tunku Abdul Rahman, Jalan Sungai Long, Bandar Sungai Long, Cheras 43000, Kajang, Selangor, Malaysia.
[2]Institute for Sustainability, Devonshire Building, Newcastle University, Newcastle Upon Tyne, NE1 7RU, U.K.
[3]Author to whom any correspondence should be addressed.
E-mail: yslim@utar.edu.my

Abstract. This project is about the development and integration of a real-time network simulator in the laboratory using hardware in the loop (HIL) for the purpose of frequency regulation. Frequency regulation is done using the energy storage system (ESS) and a real-time network test bed developed in the smart energy laboratory in Newcastle University. An IEEE Test System was built in the OPAL-RT network simulator to mimic the power grid with renewable energy sources. The study demonstrates the viability of using an ESS to regulate the frequency under an increased penetration of renewable energy sources.

1. Introduction

Frequency is a very important parameter to be regulated at 50 Hz or 60 Hz at all times in an AC electric network. Any deviation of frequency from the nominal value is caused by the imbalance between generation and demand. An excess in the demand yields the decrease in the frequency and vice versa [1]-[2]. Any significant deviation of frequency can trigger the power plants to shut down, hence causing power interruptions to the customers. Therefore, any power mismatch has to be corrected by the governors in the synchronous generators through the primary and secondary control actions. The primary control action is the first to be carried out to stabilize the frequency within seconds before the secondary action takes over to restore the frequency to the nominal value. The tertiary control will come into effect if the secondary action has not successfully restored the frequency [3].

The restoration of frequency is achieved through a sequence of control actions which may take up to several minutes. This is because the rotors in the generators have high moment of inertia that slows down the response of the generators towards any frequency changes [4]-[6]. As a result, the generators carry out the frequency regulation with low ramping rate in terms of power supply.

At present, this mechanism of frequency regulation may not pose any major concern to the utility companies or network operators because the frequency is not easily changed and the magnitudes of changes are not significant on the networks where the total inertia of the generators is still high.

However, with the increased penetration of the renewable energy sources on the networks, some generators will be replaced by renewable energy sources such as wind farms or solar farms for power generation, hence reducing the total inertia of the networks. The frequency can then be volatile and susceptible to the power mismatches. In addition, the power outputs of the wind or solar generating systems are usually intermittent, hence increasing the occurrence of power mismatches and hence the occurrence of frequency deviations.

As such, additional standby generators may be required to manage a large number of frequency changes. However, this approach can increase the cost of ancillary service to be borne by the consumers through the increase in the electricity tariffs [7]-[9]. Furthermore, the generators with low ramping rates may not be an effective means to regulate the frequency on the low-inertia networks.

Energy storage system (ESS) can be used as an effective means for frequency regulation on the low-inertia networks because it has high ramping rate that allows ESS to respond to frequency changes much faster than that of the generators.

Based on the literature review, most of the authors focused mainly on the control performance of the ESS for frequency regulation without considering how it can continuously regulate the frequency under a narrow window of state of charge (SOC). Also, the authors have not considered the best locations of the networks that ESS should be placed in order to avoid any voltage regulation issues. Apart from that, the effects of reducing the capacity of ESS on the quality of frequency regulation should be studied in order to determine the optimum capacity of ESS, so as to achieve a balance between the quality of frequency regulation and the cost of investment.

2. Methodology

2.1. Experimental model

A 4GW network model in Figure 1 was developed using Matlab Simulink as illustrated in Figure 2. Using the OPAL-RT network simulator in the laboratory, the model was run in real-time to mimic an actual network. The network is split into 3 processes in order to prevent overruns during the simulation. When a frequency deviation occurs, the model detects and reacts to the changes.

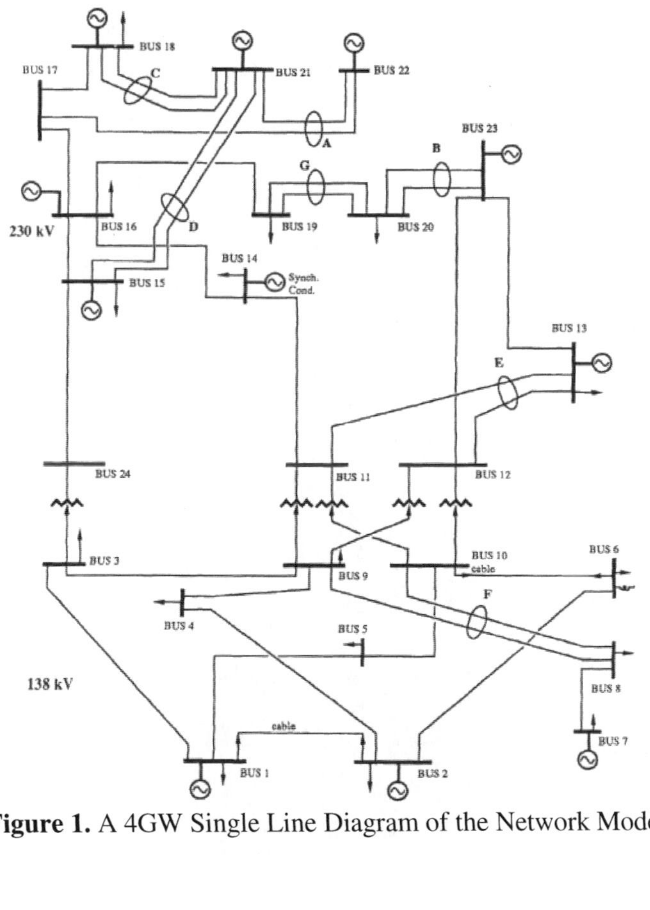

Figure 1. A 4GW Single Line Diagram of the Network Model

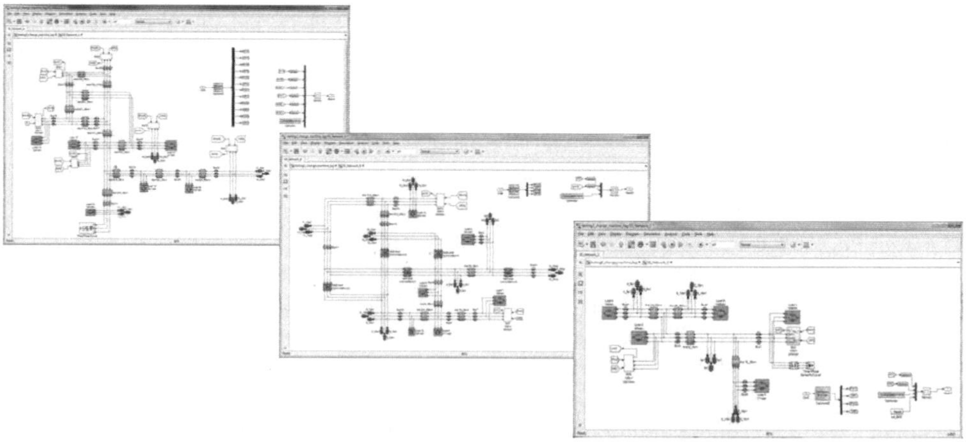

Figure 2. Model of the Network in Simulink.

A 15 kW Triphase AC/AC converter (Triphase Grid) shown in Figure 3 is used to act as the grid generation. It can be controlled in real-time to output the desired voltage magnitude and frequency, converting the signals sent from the OPAL real-time network simulator into actual voltage and frequency values on the 415 V isolated grid of the laboratory.
The 15 kW Triphase DC/AC converter (Triphase ESS) of Figure 4(a) is used to convert the DC source from the 30kW GSS Regatron battery emulator shown in Figure 4(b) to AC power and vice versa. Being a fast-response switching power electronics device, it is capable of reacting instantaneously to the deviations in frequency.
A PI controller as shown in Figure 5 is used to control the response and output of the ESS. In reality, not all the ESSs have the same controller and response time. In the design, a different slew rate (rate limiter) was added to have a different response time as simultaneous output power from multiple ESSs may create a big power surge to the network, causing instability. A lead acid battery was modelled and used in the experiments.
The hardware in the loop (HIL) was done by a dynamic load named as 'power change' in the model. Any power changes in the actual 415 V isolated network will be reflected in the network model of the OPAL-RT simulator. There are controllable loads in the model to emulate the load demand in the network. In addition, 3 photovoltaic (PV) generators were connected to the network to emulate the future grid where PV panels are installed. The overview of the laboratory integration is shown in the Figure 6.

Figure 3. Triphase Grid.

Figure 4. (a) Triphase ESS.

Figure 4. (b) Battery Emulator.

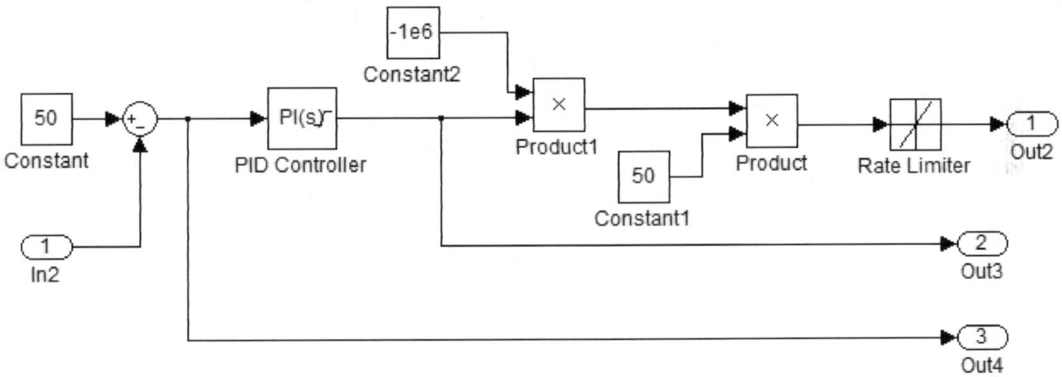

Figure 5. Design of PI Controller.

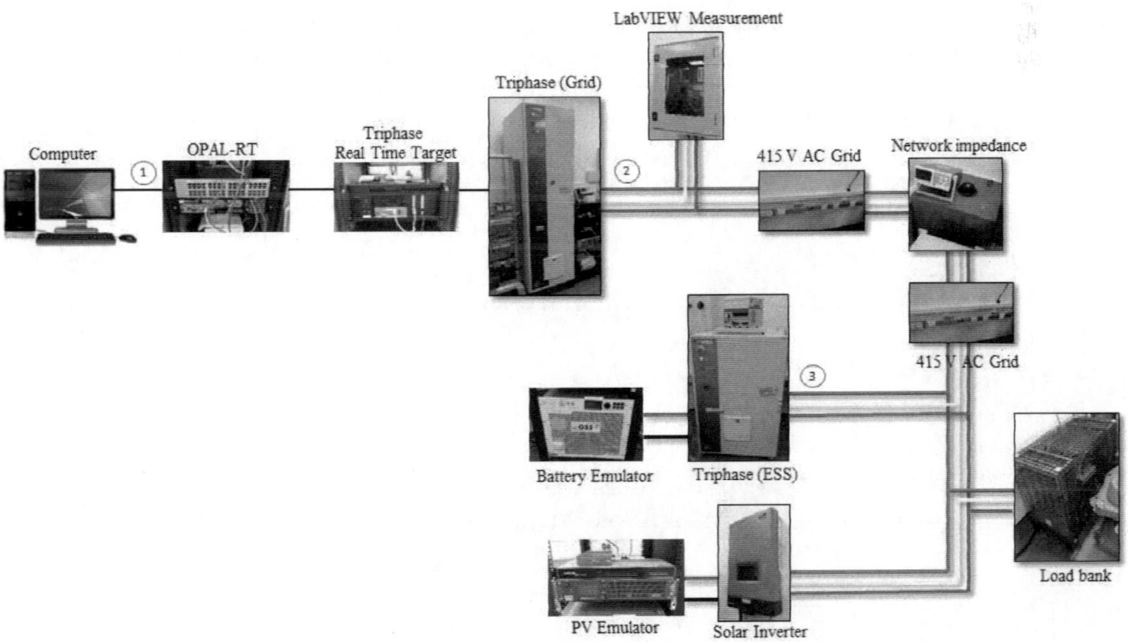

Figure 6. Overview of Laboratory Integration.

3. Results

3.1. Experimental model
Several scenarios have been depicted to verify the practicability of the ESS.
3.1.1. Case Study 1: Experiment with network simulator
IEEE 24-bus network is used in the simulation. There are 17 customer loads in the network. 4 of them are static loads while 13 of them have dynamic load profiles (30 seconds per sample). The total load demand in the network is 2850 MW while the total generation is 4037 MW. Therefore, the total loading of the network is 70.6%. PVs are introduced in different buses in the network to demonstrate the future scenario. 10%, 20%, 40%, 60%, 80%, and 100% of PVs are introduced in the experiments with the different PV profiles of 30 seconds per sample. Simulation was done using the real-time network simulator for a duration of 350 seconds (around 5 minutes) for each of the simulation. The results obtained were plotted in Figure 7.

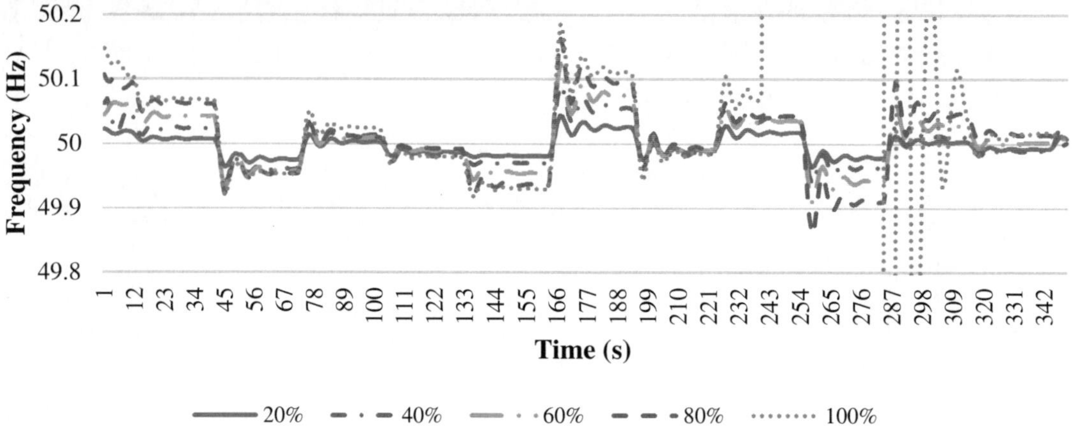

Figure 7. Frequency fluctuations with the different percentage of PVs are introduced in the 24-bus network.

When there is 100% PV in the network, the network will be disconnected due to the instability of the network. The ESS was then introduced to the network. 20% of ESS was installed when 20% of PVs are integrated in the network in order to compensate the excess power flow in the network, similarly for 40%, 60%, 80% and 100%. The results obtained are shown in Figure 8.

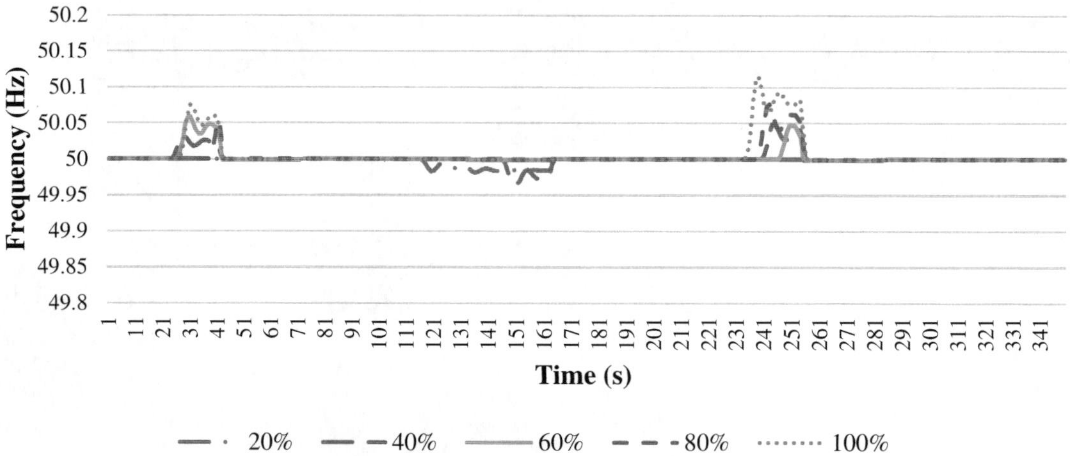

Figure 8. ESS is integrated with the ratio of 1:1 to the PVs integration in the network.

From the results, it can be seen that the fluctuation of the frequency is reduced tremendously. However, there are still some minor fluctuations due to the limited capacity of the ESS. From this, we can see that the ESS can improve the stability of the network where PVs are integrated to the network in the future.

3.1.2. Case study 2: Experiment with the laboratory setup

With the same 24-bus network used in the first case study, experiments were done using actual equipment in the laboratory. The experiment is on regulating the frequency in the event of a 50 MW and 200 MW load change using the energy storage in the laboratory. The fluctuations in the frequency are reduced with the ESS when the 50 MW and 200 MW loads were turned on and off as illustrated in Figures 9(a) and 9(b).

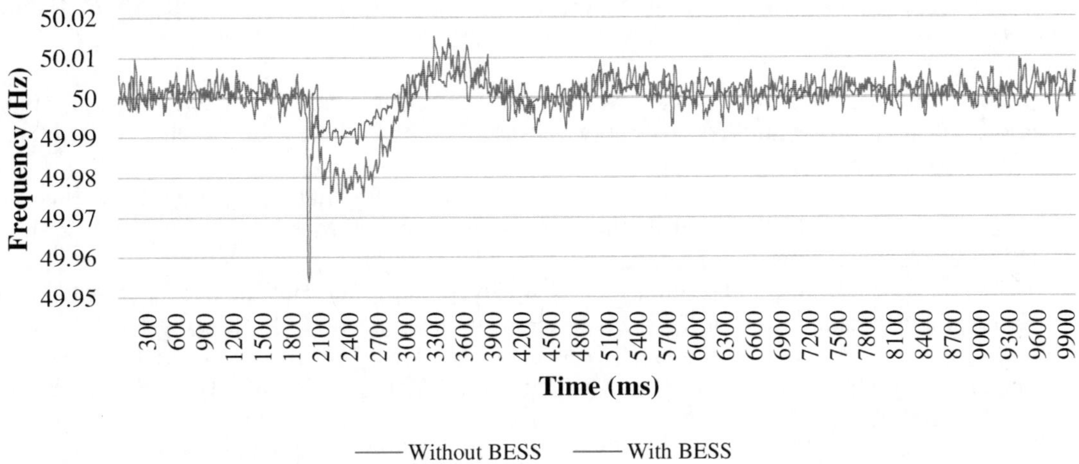

Figure 9. (a) A 50 MW Event.

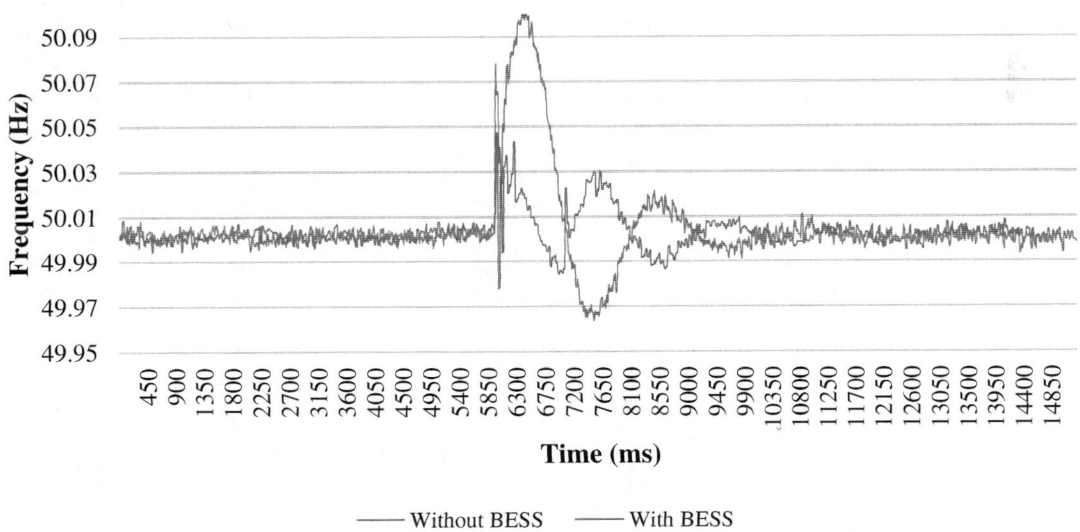

Figure 9. (b) A 200MW Event.

4. Conclusion

The ESS effectively mitigates the frequency fluctuation as demonstrated in the project. The laboratory in which the experiment was carried out can be integrated with different smart appliances where they can be used with different applications and research testing in the future. An important achievement of the project is that it can be extended for future work. Besides, HIL can solve the limitations of the hardware and software by using them together. More and more ESS applications will be developed and enhanced in the near future as ESSs become integrated into the grid to counter the intermittent output of renewables which form part of the carbon reduction plan. Besides setting up an environment for future testing, the laboratory also provides a safe environment for engineers to test out their ideas without being exposed to hazards on the actual grid. Lastly, this project also proves that the ESS is feasible to be used in frequency regulation and may help to reduce dependence on excess rotating capacity employed in conventional strategies leading to a reduction in carbon emissions.

5. Acknowledgements

The research leading to these results has received funding from the Newton Fund project in collaboration between the University of Newcastle and the Universiti of Tunku Abdul Rahman (UTAR) Malaysia.

6. References

[1] S. M. Drouilher. 1999 Power Flow Management in a High Penetration Wind-Diesel Hybrid Power System with Short-Term Energy Storage National Renewable Energy Laboratory – Diesel electric power-plants.

[2] M. L. Lazarewicz and A. Rojas 2004 Grid Frequency Regulation by Recycling Electrical Energy in Flywheels IEEE Power Engineering Society General Meeting 2004 pp 2038–2042

[3] UCTE 2004 UCTE Operation Handbook – Appendix 1 UCTE

[4] N. Jaleeli, D. N. Ewart, L. H. Fink, L. S. Van Slyck and A. G. Hoffman 1992 Understanding Automatic Generation Control IEEE Transactions Power Systems 7(3) pp 1106–1122

[5] S. A. Lefton P. M. Besuner and G. P. Grimsrud 1994 Managing Utility Power Plant Assets to Economically Optimize Power Plant Cycling Costs, Life, and Reliability EPRI Fossil Plant Cycling Conference

[6] Y. V. Makarov, J. Ma, S. Lu and T. B. Nguyen 2008 Assessing the Value of Regulation Resources based on their Time Response Characteristics Pacific Northwest National Laboratory

[7] D. C. Prowse. 1993 Improvements to a Standard Automatic Generation Control Filter Algorithm IEEE Transactions Power Systems 8(3) pp 1204–1210

[8] Erinmez I. A, D. O. Bickers, G. F. Wood and W. W. Huang. 1999 NGC Experience with Frequency Control in England and Wales – Provision of Frequency Response by Generators IEEE PES Winter Meeting

[9] S. L. Koh, Y. S. Lim and S. Morris. 2011 Potential of Advanced Coal and Gas Combustion Technologies in GHG Emission Reduction in Developing Countries from Technical, Environmental and Economic Perspectives Energy Procedia 12 pp 878–885

IOP Conf. Series: Earth and Environmental Science **154** (2018) 012013 doi:10.1088/1755-1315/154/1/012013

Analysis of turbine-grid interaction of grid-connected wind turbine using HHT

A Chen[1], W Wu[1], J Miao[1] and D Xie[1]

[1] Department of Electronic Information and Electrical Engineering, Shanghai Jiao Tong University, 800 Dongchuan Road, Shanghai, China

Abstract. This paper processes the output power of the grid-connected wind turbine with the denoising and extracting method based on Hilbert Huang transform (HHT) to discuss the turbine-grid interaction. At first, the detailed Empirical Mode Decomposition (EMD) and the Hilbert Transform (HT) are introduced. Then, on the premise of decomposing the output power of the grid-connected wind turbine into a series of Intrinsic Mode Functions (IMFs), energy ratio and power volatility are calculated to detect the unessential components. Meanwhile, combined with vibration function of turbine-grid interaction, data fitting of instantaneous amplitude and phase of each IMF is implemented to extract characteristic parameters of different interactions. Finally, utilizing measured data of actual parallel-operated wind turbines in China, this work accurately obtains the characteristic parameters of turbine-grid interaction of grid-connected wind turbine.

1. Introduction

Currently, with the increasing permeability of wind power in power grid, more and more remarkable environmental and economic benefits are achieved. However, problems of power-quality caused by grid-connected wind turbines should not be ignored [1]. Due to the interactions in components of electrical coupling and mechanical connection between wind turbines and power grid, output waveforms of wind turbines contain various kinds of vibration signals with different frequencies, which are referred to as turbine-grid interaction, during the grid-connected operations [2]. At present, turbine-grid interaction is commonly classified into two categories: sub-synchronous interaction (SSI) [3] and low-frequency oscillation (LFO) [4]. Then, SSI is divided into sub-synchronous control interaction (SSCI) and sub-synchronous torsional interaction (SSTI), which is composed of sub-synchronous resonance (SSR) and sub-synchronous oscillation (SSO). In figure 1, the relationship of them is clearly illustrated.

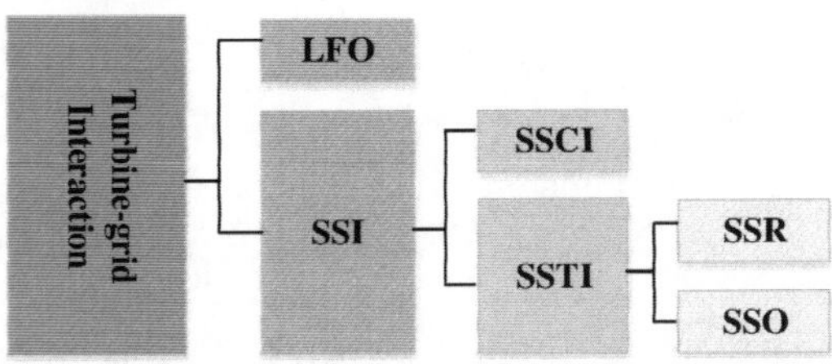

Figure 1. classification diagram of turbine-grid interaction

The turbine-grid interaction can result in reduced efficiency, increased life loss of wind turbines and rising failure rate of sensitive equipment, leading to sharp fall in productivity and influencing the security and stability of power grid operation [5], [6]. Aiming at the destructive effects of the turbine-gird interactions on the normal operation of power grid, a plenty of studies focus on the methods to identify accurately the interactions implicit in the output waveforms of wind turbines, which lay a foundation for researches on corresponding countermeasures that can reduce operational and financial damage caused by turbine-grid interactions. The main methods are twofold. The first method is small signal analysis, whose weaknesses are the construction of complex small perturbation model and off-line application [7]. The other method is modern signal processing, including fast Fourier transform (FFT) [8], wavelet transform (WT) [9] and, recently, Hilbert Huang transform (HHT) [10].

Since HHT can adaptively decompose complicated waveforms with an excellent time resolution (the time at which frequencies change), it is suitable for analysing nonlinear and nonstationary data detected in power system. In [11], a filtering method based on HHT is proposed to extract low frequency oscillation in Western Japan 60-Hz power system. Authors of [12] present a modified HHT for time-frequency analysis of distorted power quality signals.

This paper exploits the potential of HHT to implement noise detection and reduction of output power of grid-connected wind turbines and extract characteristic parameters of turbine-grid interaction. The following structure is as follows: Section 2 introduce the principle of HHT, and the extraction flow is presented in section 3. In section 4, there is a case analysis. Lastly, conclusions are drawn in section 5.

2. HHT

Output waveforms of wind turbines are typically nonlinear and nonstationary signals, so that HHT is well suited for processing them. HHT consists of two steps: EMD and HT [12].

2.1. EMD

The essence of EMD is decomposing the original signal into a series of Intrinsic Mode Functions (IMFs), which is axial symmetry, and a residue, as shown in (1).

$$x(t)=\sum_{i=1}^{n} I_i(t)+r_n(t) \tag{1}$$

Where $x(t)$ is the original signal, $I_i(t)$ is the ith IMF, $r_n(t)$ is the residue that represents the average trend of the original signal.

Simultaneously, IMF must meet the succeeding conditions: 1) There is one zero crossing point between any adjacent local extreme point. 2) The local mean at any time is zero. Thus, the steps of EMD can be described as follows:

Step I Employ spline interpolations to obtain maxima and minima envelopes ($E_{\max}(t)$ and $E_{\min(t)}$) of the original signal $x(t)$.

Step II Calculate the candidate for the first IMF with the subsequent equation.

$$P_1=x(t)-\left[E_{\max}(t)+E_{\min}(t)\right]/2 \tag{2}$$

Step III Judge whether $P_1(t)$ meets the aforementioned requirements to be an IMF. If yes, the first IMF $I_1(t)=P_1(t)$, if not, then, repeat Step I and II on the premise of regarding $P_1(t)$ as the original signal until $P_1(t)$ becomes an IMF.

Step IV Calculate the first residue $r_1(t)$ and test whether the number of extreme point of $r_1(t)$ is less than or equal to one. If yes, terminate EMD, otherwise repeat Step I, II and III on $r_1(t)$.

2.2. HT

Once all the IMFs are extracted from the original signal, IMFs are processed by HT to obtain their instantaneous amplitudes and frequencies. The HT $\hat{I}(t)$ of the IMF signal $I(t)$ can be obtained as

$$\hat{I}(t)=H\left[I(t)\right]=\frac{1}{\pi}\int_{-\infty}^{+\infty}\frac{I(\tau)}{t-\tau}d\tau \tag{3}$$

So analytic signal of HT and its corresponding calculations are shown as follows.

$$Z(t)=I(t)+i\hat{I}(t)=a(t)\exp\left(i\theta(t)\right) \tag{4}$$

IOP Conf. Series: Earth and Environmental Science **154** (2018) 012013 doi:10.1088/1755-1315/154/1/012013

$$a(t)=\left[I(t)^2+\hat{I}(t)^2\right]^{1/2} \tag{5}$$

$$\theta(t)=\arctan\left[\hat{I}(t)/I(t)\right] \tag{6}$$

$$f(t)=\frac{1}{2\pi}\frac{d\theta(t)}{dt} \tag{7}$$

Where $Z(t)$ is analytic signal, $a(t)$ is instantaneous amplitude, $\theta(t)$ is instantaneous phase and $f(t)$ represents instantaneous frequency.

3. Processing methods

3.1. Signal denoising

For better analysis of original signal, the noise implicit in the signal should be eliminated. In this work, two indexes are employed to evaluate each IMF: energy ratio, the smaller it is, the less important the IMF is, and power volatility, the drastic change of which represents the huge role played by the IMF. IMFs which feature different amplitudes contain distinct amounts of energy, and energy of the ith IMF can be calculated as

$$W_i=\sum_{i=1}^{N}\left|a_i(t)\right|^2\Big/N \tag{8}$$

Where N is the total number of data points. Thus, energy ratio is defined as

$$R_i=W_i/W=W_i\Big/\sum_{i=1}^{n}W_i \tag{9}$$

Where W is the sum of energy. Moreover, power volatility reflects the fluctuation level of output power of wind turbines. Hence, the volatility of each IMF, which has specific frequency, can be calculated by (10) and utilized to investigate the main frequencies of wind power fluctuation.

$$V(t)=\frac{\max\left[P_{local}(t)\right]-\min\left[P_{local}(t)\right]}{P_{rated}}\times 100\% \tag{10}$$

Where $V(t)$ is the power volatility, $\max\left[P_{local}(t)\right]$ is the maximum power in a certain period of time, $\min\left[P_{local}(t)\right]$ is the minimum power, and P_{rated} is the rated capacity of wind turbine.

Extraction of characteristic parameters of turbine-grid interactions

In power system, turbine-grid interaction can be expressed as

$$x(t)=Ae^{-\lambda t}\cos\left(wt+\theta_0\right) \tag{11}$$

Where A is the initial amplitude, λ is attenuation coefficient, w is frequency and θ_0 is initial phase. When HT is carried out on the turbine-gird interaction $x(t)$, combined with (4)~(6), (11) can be transformed to the following equations.

$$a(t)=\left[x(t)^2+\hat{x}(t)^2\right]^{1/2}=Ae^{-\lambda t}\Rightarrow \ln a(t)=-\lambda t+\ln A \tag{12}$$

$$\theta(t)=wt+\theta_0 \tag{13}$$

Subsequently, data fitting of instantaneous amplitude and phase of each IMF is performed to obtain accurate A, λ, w and θ_0. Then, damping ratio can be calculated as

$$\zeta=\lambda\Big/\left[(w)^2+(\lambda)^2\right] \tag{14}$$

4. Cases analyses

To study the turbine-grid interaction between grid-connected wind power and power grid, measured data of actual parallel-operated wind turbine in China is used. Based on the above signal processing methods, the concrete analysis flow is shown in figure 2.

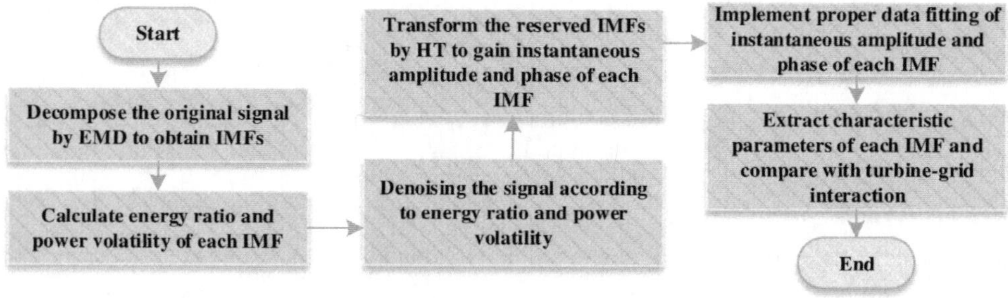

Figure 2. Analysis flow of measured data of wind turbines

The capacity of the wind turbine is 2 MW. The sampling frequency of data is 4000 Hz, and this paper takes a set of 40000 points, that is a group of 10 s data.

Firstly, the output power is decomposed into 13 IMFs and a residue, as shown in figure 3. Then we can gain instantaneous amplitude and phase and calculate the energy ratio and power volatility of each IMF, which can be clearly found in table 1 and figure 4. Obviously, energy ratios of IMF8~IMF13 is very small and power volatilities of IMF5~IMF13 keep basically stable, which mean that IMF8~IMF13 have small fluctuation and include little information of output power. Thus, the elimination of these components has slight impact on the analysis of the signal.

Table 1. Energy ratio and power volatility (%)

IMF	1	2	3	4	5	6	7	8	9	10	11	12	13
Energy ratio	58.51	12.33	13.69	8.88	3.30	1.68	0.86	0.36	0.17	0.10	0.06	0.02	0.03

According to the expression of vibration function (11) we can obtain A, λ, w and θ_0 by fitting instantaneous amplitude and phase of IMF1~IMF7, as shown in table 2. And the approximate frequencies are 48.46 Hz, 21.67 Hz, 14.32 Hz, 6.01 Hz, 2.01 Hz, 1.62 Hz and 0.37 Hz. Compared with the results of small signal analysis on the same wind turbine, we can find that 21.67 Hz is the frequency of SSR, 14.32 Hz, 2.01 Hz and 1.62 Hz are the frequencies of SSO, 6.01 Hz is the frequency of SSCI, 0.37 Hz is the frequency of LFO. Also, based on the initial amplitude we can conclude that the influence of SSCI is greater than SSR and LOF. And SSO is common in the output power. Additionally, ζ of each IMF can be calculated by (14).

Table 2. Characteristic parameters

IMF	1	2	3	4	5	6	7
A	3.999e4	1.693e3	1.833e4	1.474e4	8.418e3	6.275e3	4.297e3
λ	0.0046	0.0015	0.0026	0.0082	0.0029	0.0062	0.0018
w	304.313	136.096	89.938	37.736	12.623	10.167	2.333
θ_0	-0.0034	0.0036	0.0079	0.0064	-0.0011	0.0006	0.0069
ζ	0.212	0.672	0.034	0.721	0.344	0.362	0.325

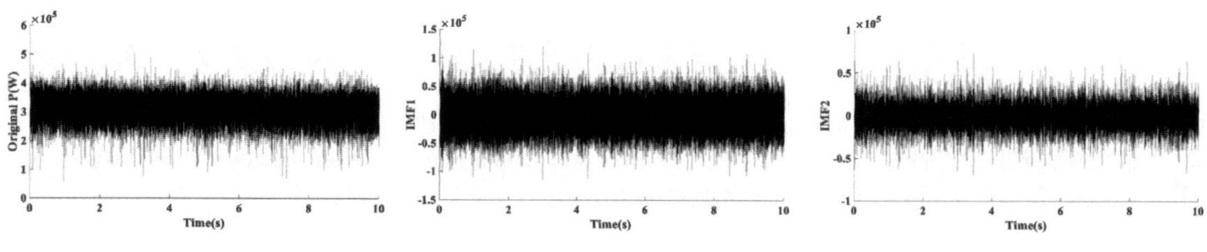

IOP Conf. Series: Earth and Environmental Science **154** (2018) 012013 doi:10.1088/1755-1315/154/1/012013

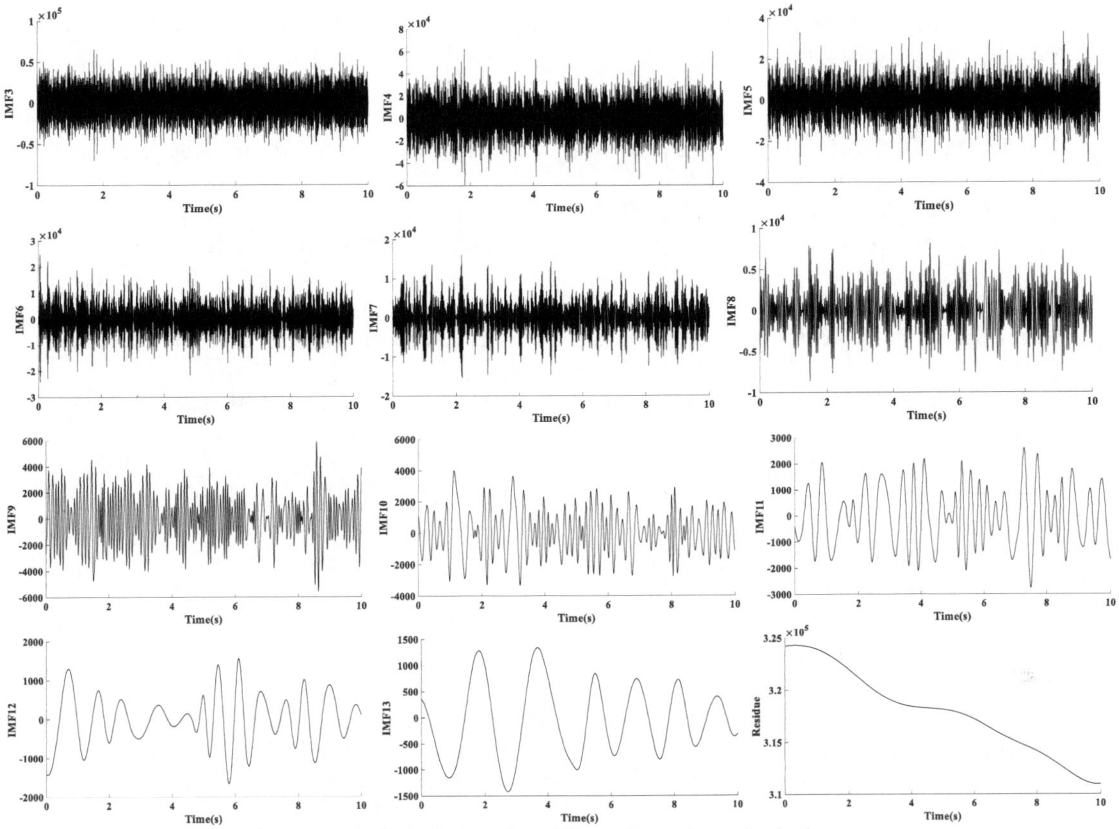

Figure 3. Original signal, IMFs and residue of wind power

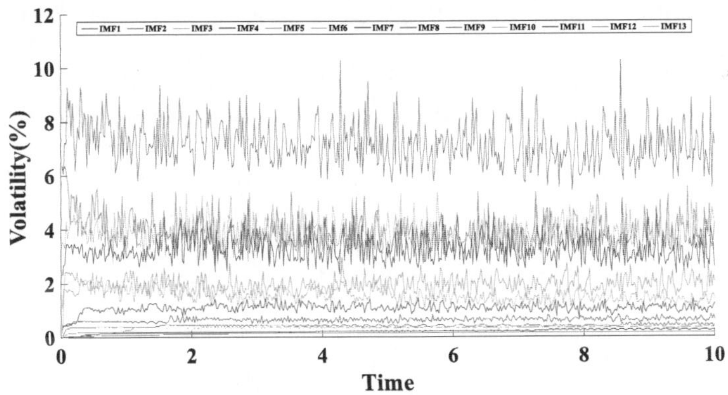

Figure 4. Power volatility of IMFs

5. Conclusions

This paper effectively analyses turbine-grid interaction in output power of wind turbine with methods based on HHT. And with the results of actual case, the following conclusions can be drawn:

- Low frequency noise can be removed from the output power of wind turbine according to energy ratios and power volatilities after the EMD process.
- In output power of wind turbine, SSR, SSO, SSCI and LOF can be find clearly, and they all have impact on the output power.
- Specifically, the influences of SSCI and SSO are severer than SSR and LOF. And SSO with the frequency of 14.32 Hz has a stronger effect than 2.01 Hz and 1.62 Hz.

Since the ability of HHT drop off rapidly when it is used to process the signal containing components with close frequencies, results obtained by methods based on HHT always have some error. Thus, we can improve the skill of frequency resolution of our methods to polish up the case results and further study the turbine-grid interaction.

6. References

[1] A. S. Bubshait, A. Mortezaei, M. Simoes, et al. Power Quality Enhancement for a Grid Connected Wind Turbine Energy System [J]. IEEE Transactions on Industry Applications, 2017, PP (99):1-1.

[2] B. Badrzadeh, M. Sahni, Y. Zhou, et al. General methodology for analysis of sub-synchronous interaction in wind power plants [C]// Power and Energy Society General Meeting. IEEE, 2013:1858 - 1869.

[3] H. Liu, X. Xie, J. He, et al. Subsynchronous Interaction between Direct-Drive PMSG Based Wind Farms and Weak AC Networks [J]. IEEE Transactions on Power Systems, 2017, PP (99):1-1.

[4] T. M. Cesar, S. P. Pimentel, E. G. Marra, et al. Wavelet Transform analysis for grid-connected photovoltaic systems [C]// International Conference on Clean Electrical Power. 2017:1-6.

[5] M. Ghafouri, U. Karaagac, H. Karimi, et al. An LQR Controller for Damping of Subsynchronous Interaction in DFIG-Based Wind Farms [J]. IEEE Trans. on Power Systems, 2017, PP (99):1-1.

[6] C. Li, W. Zhang, R. Liu. Forced low frequency oscillation of wind-integrated power systems [C]// Innovative Smart Grid Technologies Conference. IEEE, 2016:1-5.

[7] H. Liu, X. Xie, J. He, et al. Subsynchronous Interaction between Direct-Drive PMSG Based Wind Farms and Weak AC Networks [J]. IEEE Transactions on Power Systems, 2017, PP (99):1-1.

[8] L. Peretti, M. Pathmanathan, O. I. U. Haq, et al. Robust harmonic detection, classification and compensation method for electric drives based on the sparse fast Fourier transform and the Mahalanobis distance [J]. Iet Electric Power Applications, 2017, 11(7):1177-1186.

[9] Z. L. Zhang , X. Cheng, Z. Y. Lu, et al. SOC Estimation of Lithium-ion Batteries with AEKF and Wavelet Transform Matrix [J]. IEEE Transactions on Power Electronics, 2016, PP (99):1-1.

[10] A. Gururani, S. R. Mohanty, J. C. Mohanta. Microgrid protection using Hilbert–Huang transform based-differential scheme [J]. Iet Generation Transmission & Distribution, 2016, 10 (15):3707-3716.

[11] Q. Liu, Y. Mitani. Application of HHT for oscillation mode analysis in power system based on PMU[C]// International Conference on Electric Power and Energy Conversion Systems. IEEE, 2012:1-4.

[12] M. J. Afroni, D. Sutanto, D. Stirling, Analysis of Nonstationary Power-Quality Waveforms Using Iterative Hilbert Huang Transform and SAX Algorithm [J]. IEEE Transactions on Power Delivery, 2013, 28(4):2134-2144.

IOP Conf. Series: Earth and Environmental Science 154 (2018) 012014 doi:10.1088/1755-1315/154/1/012014

Simulation of load-sharing in standalone distributed generation system

Titus O Ajewole[1], Robert P M Craven[2], Olakunle Kayode[3] and Olufisayo S Babalola[4]

[1]Department of Electrical and Electronic Engineering, Osun State University, Osogbo, Nigeria
[2]Centre for Energy Systems Research, Tennessee Technological University, Cookeville, USA
[3]Department of Mechanical Engineering, Osun State University, Osogbo, Nigeria
[4]Department of Electronic and Electrical Engineering, Obafemi Awolowo University, Ile-Ife, Nigeria

Abstract. This paper presents a study on load-sharing among the component generating units of a multi-source electric microgrid that is operated as an autonomous ac supply-mode system. Emerging trend in power system development permits deployment of microgrids for standalone or stand-by applications, thereby requiring active- and reactive power sharing among the discrete generating units contained in hybrid-source microgrids. In this study, therefore, a laboratory-scale model of a microgrid energized with three renewable energy-based sources is employed as a simulation platform to investigate power sharing among the power-generating units. Each source is represented by a source emulator that captures the real operational characteristics of the mimicked generating unit and, with implementation of real-life weather data and load profiles on the model; the sharing of the load among the generating units is investigated. There is a proportionate generation of power by the three source emulators, with their frequencies perfectly synchronized at the point of common coupling as a result of balance flow of power among them. This hybrid topology of renewable energy-based microgrid could therefore be seamlessly adapted into national energy mix by the indigenous electric utility providers in Nigeria.

1. Introduction

There is currently a prevalent of energy deficit across sub-Saharan Africa where close to 70 % of the population general do not have access to electricity, while Nigeria in particular has 40% of its population living without electric power [1]. At present, gas-fired plants accounts for not less than 64% of the electricity generation in Nigeria, with most of the power system infrastructures currently operated near their steady-state stability limits. According to the Energy Commission of Nigeria (ECN), the current generation capacity of the country is about 4,000 MW for a population of approximately 140 million [2]. As predicted by the ECN, demand for electric power in Nigeria may increases to hit 250,000 MW by year 2030 and so there is an urgent need for improvement on the existing generation mix in order to meet both the current and the projected demands. Any further expansion would definitely require huge investments in plant infrastructures, such that could be very difficult to achieve in the face of the prevailing national economic downturn. However, one realistic way to tackle this challenge is to increase the renewable energy quota in the national power generation mix [3], [4], since Nigeria is abundantly endowed with renewable energy resources that are yet to be adequately harnessed [5], [6].

With regards to the emerging trend in power system development across the globe, renewable energy-based electric microgrids are being employed for stand-alone, stand-by, grid-connected, co-generation or peak-shaving applications and to ensure a seamless adoption of renewable energy technology (RET) into the Nigerian electric power sub-sector, the place of research and development is very critical in the exploration of the local energy resources. Energy development requires on-going research into understanding new systems, including renewable energy systems [7] and, according to [8], [9], reduced-scale models of renewable energy technology systems could be used to understudy the behaviours of the actual systems in the real-life scenario.

In the drive towards renewable energies, reliably balancing load and demand is important as the share of renewables in the energy mix increases in the growing importance of distributed generation (DG). In this study therefore, load-sharing among the component generating units of a microgrid system with multiple energy sources, which is operated in standalone ac supply-mode is investigated using a laboratory-scale model (lab-grid) of the system. Load-sharing is very crucial in distributed power systems because real- and reactive power controls must operate independently of each other and share a common real- and reactive load in proportion to a pre-determined ratio regardless of plant parameters [10].

Load-sharing in distributed generation systems has been investigated by various authors. Small-signal stability analysis of combined droop and average power method for load sharing control of multiple distributed generation systems in a standalone ac supply mode is discussed in [11] where the authors show that the small-signal model accurately predicts the stability of the parallel systems. The focus in [10] is on improvement to the conventional frequency droop technique of real power-sharing and also to develop a new reactive power-sharing scheme. The improvement allows the operator to tune the real power-sharing controller to achieve desired system response without compromising frequency regulation. The author in [12] studies the flow of active- and reactive powers in two distributed generators that has common load but with different control functions. The simulation is done with PSCAD software and voltage source converter is employed for compensation. In [13], droop controller is analyzed and optimized based on small-signal dynamic modelling.

In this study, a laboratory-scale model of a multiple-energy-source microgrid system is employed to investigate load-sharing among the component energy sources of the system. Since cutting-edge facilities required for research in renewable energies are still lacking in most of the Nigerian tertiary institutions [6], the model is a simulation platform developed for use as an improvised teaching and research facility. The lab-grid is made of three energy source emulators that are interconnected at a point of common coupling to supply an aggregate of electrical loads. The emulators capture and mimic the operational characteristics of two pico-hydropower (PHP) sources and a wind turbine (WT) source with the three sources synchronized and monitored by the aid of a hierarchical control function.

System modelling has been a useful approach in investigating microgrid technology, yet [14] submits that models demonstrated with implementation of simulated weather data and load demand profiles may not accurately represent the real-world-scenario operational behaviours of the actual system. More so, the indigenous utility providers in Nigeria do not want to adopt any of the emerging renewable energy technologies without exhaustive validations using real-life data and profiles obtained from the local terrain.

Therefore, a real-life wind speed profile of a Nigerian locale is implemented on the WT emulator to investigate the behaviours of the lab-grid under a near-real-world operating condition. Implementation of real-life weather data and load demand profiles on system models has yielded more precise emulations of actual systems [15]-[17], therefore the load sharing among the three source emulators of the system under study is demonstrate through the real-time simulation of the system. The emulators complement one another through their proportionate generation of power, while the output frequencies of the three are perfectly synchronized as a result of the balance flow of power among the three units and, between the units and the load.

This paper is organized as follows: Section 2 of this paper describes the mathematical representation and the design structure of the lab-grid with the load-sharing control mechanism, while Section 3 presents and discusses the result obtained from the simulation and, Section 4 concludes the paper.

2. System structure

The simulation model has, by design, two PHP source emulators and a WT emulator. As shown in figure 1, one of the PHP source emulators is made to form a local grid that sets the magnitude and the frequency of the system's operating voltage. The ac voltage generated by the grid-forming source makes the reference to which the other two sources are synchronized as the two feed power to the local grid.

An important requirement of the application is that the two grid-feeding sources had to be perfectly synchronized to the local grid to keep the system voltage and frequency close to the rated values. Mechanism of the control is oriented in such a way as to minimize the operational cost of the lab-grid, while maximizing its efficiency, reliability, and controllability. Each of the source emulators is interfaced with the grid bus through a separate converter system using voltage orientation control. The control function is based on the transformation between the *abc* stationary reference frame and the *dq*-axis synchronous reference frame and the algorithm is implemented in the grid-voltage synchronous reference frame that makes all the variables to be of dc components in the steady state and, thus facilitates the control of the converters for synchronized power delivery.

Rotor speed feedback control is employed on the rotor circuit of the synchronous generator of the grid-forming source using the maximum torque per ampere technique. For a salient pole machine, the dynamic model is given by [18] as:

$$\begin{cases} v_{ds} = -R_s i_{ds} + \omega_r L_q i_{qs} - L_d p i_{ds} \\ v_{qs} = -R_s i_{qs} - \omega_r L_d i_{ds} + \omega_r \lambda_r - L_q p i_{qs} \end{cases} \tag{1}$$

$$T_e = \frac{3P}{2}\left[\lambda_r i_{qs} - (L_d - L_q)\left(i_s^2 - i_{qs}^2\right)^{-1} i_{qs}\right] \tag{2}$$

$$P_m = \frac{3}{2}\left(v_{ds} i_{ds} + v_{qs} i_{qs}\right) \tag{3}$$

$$\begin{cases} i_{qs}^* = \dfrac{2T_e^*}{3P\left[\lambda_r - (L_d - L_q)\right] i_{ds}} \\ i_{ds}^* = \dfrac{\lambda_r}{2(L_d - L_q)} \pm \left[\dfrac{\lambda_r^2}{4(L_d - L_q)^2} i_{qs}^2\right]^{-1} \end{cases} \tag{4}$$

Where v_{ds}, v_{qs}, i_{ds}, i_{qs}, are *dq*-axes stator voltages and currents respectively, i_{ds}^* and i_{qs}^* are *dq*-axes reference currents, L_d, L_q, are *dq*-axes stator self-inductances, R_s, is the stator resistance, ω_r, λ_r, are rotor electrical angular speed and rotor flux linkage respectively, P is number of pole, while T_e, P_m are the electromechanical torque and the mechanical power respectively of the synchronous machine. At the grid side, the grid-forming converter is controlled to generate a sinusoidal voltage whose magnitude and frequency is determined by the reference amplitude v^* and frequency w^*. The voltage is delivered to the lab-grid bus via the converter and the external loop of the scheme controls the grid voltage to match its reference values, while the internal control loop regulates the current supplied by the converter [19].

The power delivered to the local grid by the two grid-feeding sources is in synchronism with that of the grid-forming source. The first grid-feeder is also a PHP source emulator that is also based on wound rotor synchronous generator with rotor speed feedback control employed on the rotor circuit. The second grid-feeding source is a WT emulator of the variable-speed configuration that employs a wound rotor induction generator with reduced-capacity converter. To permit the adjustment of the amplitude and the frequency of the output voltage through reference powers P^* and Q^*, the grid-feeding interface converters are controlled using grid voltage orientation control technique.

IOP Conf. Series: Earth and Environmental Science **154** (2018) 012014 doi:10.1088/1755-1315/154/1/012014

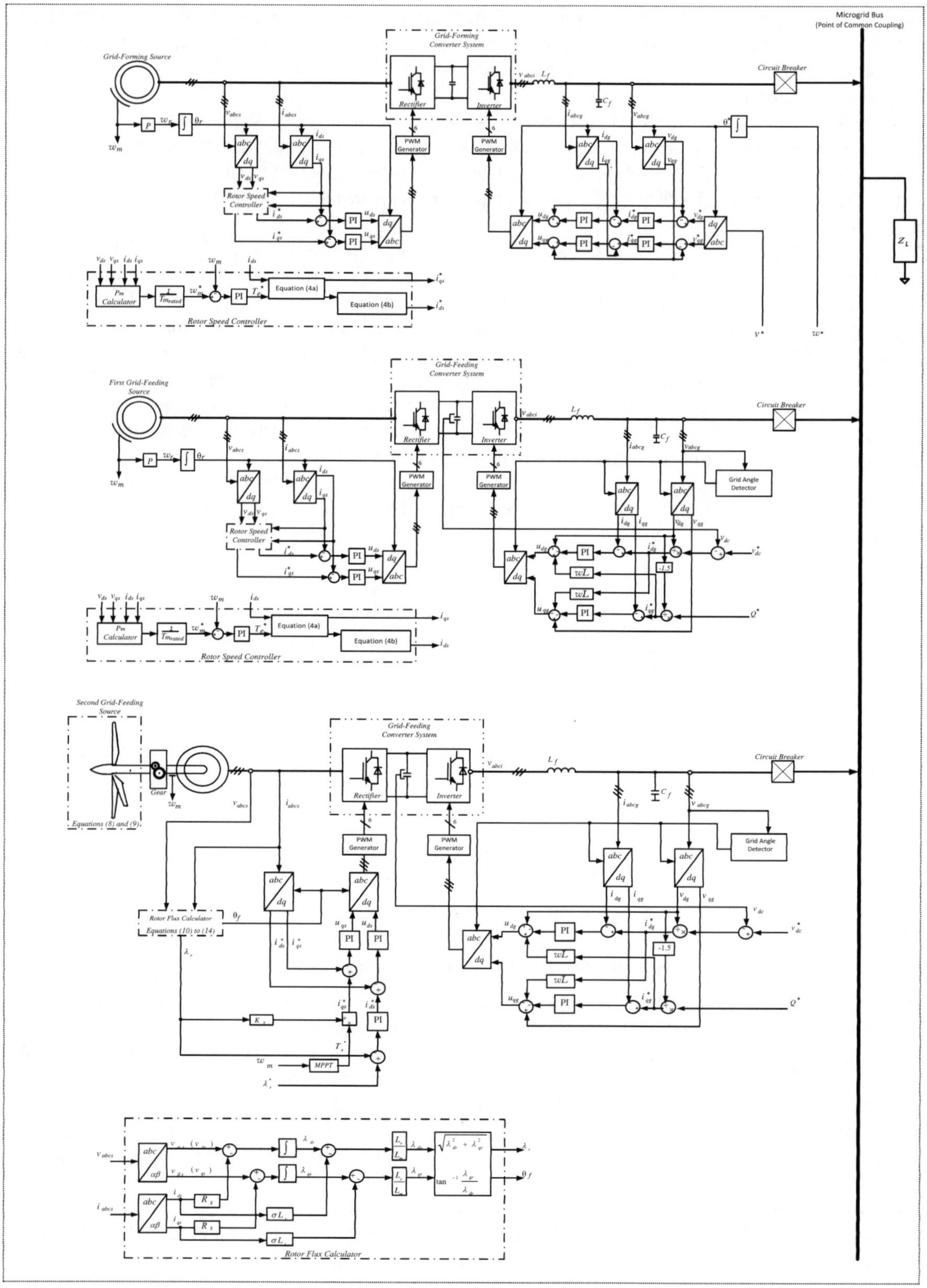

Figure 1. The stand-alone lab-grid.

IOP Conf. Series: Earth and Environmental Science **154** (2018) 012014 doi:10.1088/1755-1315/154/1/012014

$$\begin{cases} L_f pi_{dg} = v_{dg} - v_{di} + \omega_g L_f i_{qg} \\ L_f pi_{qg} = v_{qg} - v_{qi} - \omega_g L_f i_{dg} \end{cases} \tag{5}$$

For a satisfactory dynamic performance [20], a decoupled proportional-integral (PI) controller is employed, which eliminates the cross-coupled effect of the *d*- and *q*-axes currents.

$$\begin{cases} L_f pi_{dg} = K\left(i_{dg}^* - i_{dg}\right) \\ L_f pi_{qg} = K\left(i_{qg}^* - i_{qg}\right) \end{cases} \tag{6}$$

$$\frac{i_{dg}}{i_{dg}^*} = \frac{k_p p + k_I}{L_f p^2 + k_p p + k_I} \tag{7}$$

Where L_f is the inductance of the grid-interface filter, v_{dg}, v_{qg}, i_{dg}, i_{qg} are the *dq*-axes grid voltages and currents respectively, ω_g is the angular frequency of the local grid. $K=(k_p+k_i/S)$ is the gain of the controller, with k_p being the proportional gain while k_i is the integral gain.

The WT emulator is designed to have both software and hardware components. The soft component is developed on the platform of the real-time simulation computer aided design, which is a software integral of the real time digital simulator [21]. While the emulator is designed for 8m/s and 14m/s cut-in and cut-out speeds respectively, signal output of the soft component excites a direct current motor that is conditioned to operate in the torque control mode with an external analogue voltage signal. The actual electric power is generated by a wound rotor induction machine being driven at super-synchronous speed by the motor. In the approximate, the mechanical power extractable from the wind by the rotation of turbine blades is generally expressed as [22], [23].

$$P_{mech} = \frac{1}{2}\rho A C_p v^3 \tag{8}$$

Where P_{mech}, ρ, A, C_p and v are the mechanical power developed by the turbine, air density, swept area of the rotor blade, the power coefficient and wind speed respectively.

Equation (8) is modelled on the real-time simulation computer aided design platform using the component model library of the simulator, with the maximum mechanical power being

$$P_{mech_{max}} = K\omega_{t\,max}^3 \tag{9}$$

Where ω_t is the rotational speed of the turbine. The torque output of the soft model is multiplied by a gear ratio and converted to voltage signal for the excitation of the interfaced motor. Voltage signals, corresponding to the torques developed at instantaneous speeds of the wind, thus feed into the motor through the drive and the electric power is thereby generated as the motor drives a directly connected three-phase induction machine at super-synchronous speed. Grid side control of the wind turbine emulator is also based on the grid voltage orientation control technique. However, the rotor side control is achieved through direct field oriented control method for a decoupled control of the rotor flux and the electromagnetic torque in order to achieve a high dynamic performance. Using rotor flux orientation, the stator current is decomposed into a flux-producing component and a torque-producing component with the two components controlled independently [18, 20].

$$P_m = T_e \omega_m = 2P^{-1}\omega_r (1 - S) \tag{10}$$

P is the number of poles, T_e, and P_m are the electro-mechanical torque and the mechanical power respectively and, ω_m and ω_r are the rotor mechanical speed and the electrical speed respectively.

3. Simulation and demonstration

In demonstrating the load-sharing, a real-life wind speed profile is implemented on the second grid-feeder. Location of the wind capture is Obafemi Awolowo University, Ile-Ife, Nigeria (Latitude 7.5°North/Longitude 4.3°East) at a turbine height of 114 meter. Choice of the location for the wind speed profiling is premised on the desire of the university to leverage on the available renewable energy resources for an autonomous electricity generation in order to enhance the security of power supply within the campus community.

While the grid-forming and the first grid-feeding source emulators are each designed with 250W rated synchronous generators, the induction machine of the second grid-feeding source has manufacturer rating of 300W. An aggregate load of 400W is supplied by the lab-grid and, considering the intermittent nature of the wind flow, the ability of the system to satisfy the reactive power need of the second grid-feeding source, the proportional supply of active power to the load by the three energy sources, and frequency synchronism among the three generating units are investigated.

4. Discussion of result

The waveform of the real-life wind speed profile that is implemented on the second grid-feeding source, as normalized by the giga-processor card of the real time digital simulator, is shown in figure 2. On the profile is a period of time with wind speed as low as 7.5 m/s, which is below the design cut-in speed of the WT emulator. Due to the pitching mechanism at such instant of low wind speed, there is neither active power generation nor reactive power consumption by the WT emulator, as could be found on figure 3 and figure 4. Consequently, both the grid-forming source and the first grid-feeding source jointly rise to meet the active power demand of the load at that instant, with no reactive power generation by the two sources.

Figure 3 and figure 4 respectively show the reactive- and the active power flow on the entire system. It is revealed through Figure 3 that all the reactive power requirement of the second grid-feeding source, which is about 290VAR at maximum and varies with the pattern of the wind speed profile, is jointly and adequately supplied almost equally by the grid-former and the first grid-feeder. The active power need of the load is shared by the three sources as shown in the figure 4.

While the second grid-feeding source, depending on the wind speed pattern, is able to supply power up to its maximum rating of 300W, the other two sources jointly make up for the excess power required by the load. Consequent upon this balance power flow among the three micro-generating units and between the units and the load, figure 5 shows the frequency of the grid-former perfectly tracked by the two grid-feeders, and so synchronism is established among the three micro-generating sources.

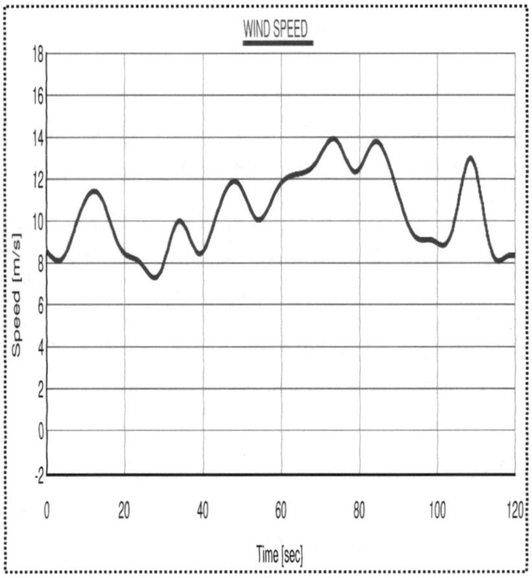

Figure 2. Real-life wind speed profile.

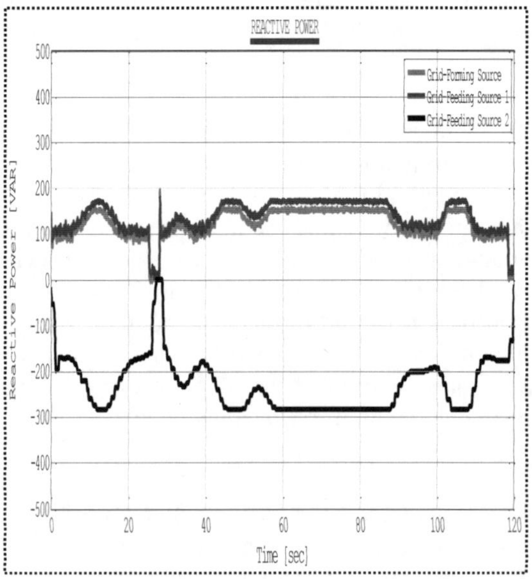

Figure 3. Reactive power flow of the system.

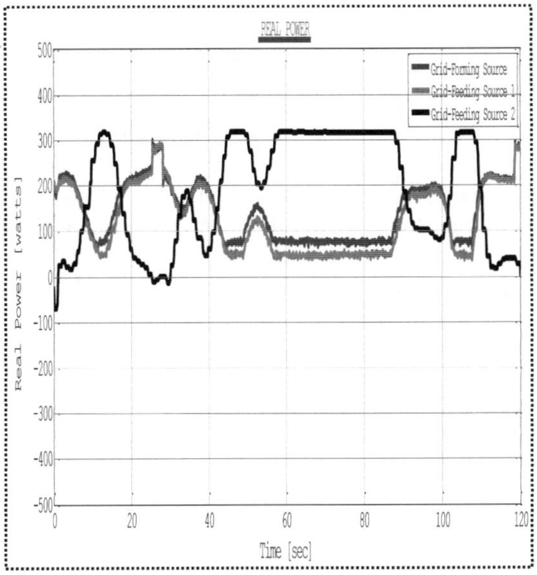

Figure 4. Active power flow of the system.

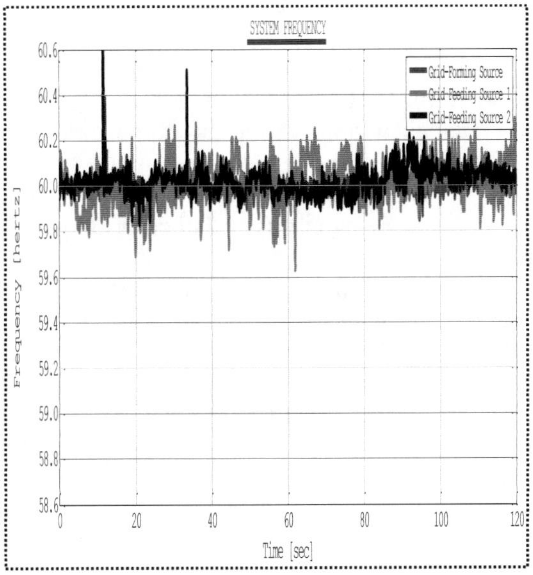

Figure 5. Frequencies of the system.

5. Conclusion

In this study, load-sharing in a hybrid-source autonomous electric microgrid is demonstrated using a system model that is developed to serve as an affordable laboratory-scale teaching and research facility on renewable energy technology. One useful approach to studying microgrid technology is the use of system emulating models. Real-life weather data and load demand profiles are implemented on a model of the microgrid for an accurate emulation of the system. There is proportionate generation of power by the component micro-generating sources of the system. The sources are able to complement each other, which is the essence of the hybridization. Also, frequencies of the three micro-generating sources are perfectly synchronized as a result of the balance power flow among the and between the units and the load. This ability to share power proportionately, and consequently a stable frequency of operation throughout the simulation period, establishes a positive load-sharing capability by the system.

Therefore, this topology of renewable energy-based hybrid-source power generation could be seamlessly adapted into the Nigerian national energy mix in the effort to alleviate the prevailing shortage of electric power in the country.

6. References

[1] B. Nnaji 2011 *Power Sector Outlook in Nigeria: Governments Renewed Priorities*. www.nigeria powerreform.org

[2] Federal Republic of Nigeria 2006 *Population and Housing Census* National Population Commission Abuja, Nigeria www.nationalpopulationcommission.ng.gov

[3] Renewable Energy Policy Network for the 21st Century 2015 *Renewables 2015 Global Status Report, Renewable Energy Policy Network for the 21st Century*. https://www.ren21.net/wp-content /uploads /2015 /07/ REN12-GSR2015

[4] International Renewable Energy Agency 2015 *Africa 2030 Roadmap for a Renewable Energy Future*. https://www.irena.org/DocumentDownloads/Publications/IRENAAfrica2030Remap2015low-res

[5] PricewaterhouseCoopers 2013 Nigeria's Third Power Sector Roundtable *A Privatised Power Sector - The Pain and the Glory*. https://www.pwc.com/ng/en/assets/pdf/pwc-a-privatized-power-sector-the-pain-and-the-glory

[6] P. C. Richard. 2016 Energy crisis in Nigeria: implication for engineering practice *NSE Tech. Trans* **50** 91 – 96
[7] A. Swart J and Hertzog P E 2016 Varying percentages of full uniform shading of a PV module in a controlled environment yields linear power reduction *J Energy in Southern Afr.* **27** 28 –38
[8] Hajimiragha A H, Dadash-Zadeh M R, Moazeni S 2015 Microgrids frequency control considerations within the framework of the optimal generation scheduling problem *IEEE Trans on Smart Grid* **6** 534 – 47
[9] J. Sachs and O. Sawodny 2016 A two-stage model predictive control strategy for economic diesel-pv-battery island microgrid operation in rural areas *IEEE Trans on Sust. Energy* **7** 903 – 13
[10] C. K. Sao and P. W. Lehn. 2005 Autonomous load-sharing of voltage source converters *IEEE Trans on Power Deliv.* **20** 1009 –16
[11] M. N. Marwali, J. W. Jung and A. Keyhani 2004 Control of distributed generation systems - part ii: load sharing control *IEEE Trans on Power Electro.* **19** 1551 – 61
[12] M. Rezaei. 2008 Load-sharing between two DGs and control of active and reactive power with vsc and optimization of vsc's parameters. Accessed 4 Dec 2014
[13] K. Yu, Q. Ai, S. Wang, J. Ni and T. Lv 2016 Analysis and optimization of droop controller for microgrid system based on small-signal dynamic model *IEEE Trans on Smart Grid* **7** 695 – 705
[14] P. M. O. Gebraad, F. W. Teeuwisse, J. W. Van-Wingerden, P. A. Fleming, S. D. Reuben, J. R. Marden and L. Y. Pao. 2016 Wind plant power optimization through yaw control using a parametric model for wake effects – A CFD simulation study*Wind Energy* **19** 95 –114
[15] C. Wang and H. M. Nehrir. 2008 Power management of a stand-alone wind/photovoltaic/fuel cell energy system *IEEE Trans on Energy Conv.* **23** 957– 67
[16] M. Asmine, J. Brochu, J. Fortmann, R. Gagnon, Y. Kazachkov, C. E. Langlois, C. Larose, E. Muljadi, J. MacDowell, P. Pourbeick, S. E. Seman and K. Wiens 2011 Model validation for wind turbine models *IEEE Trans on Power Syst.* **26** 1776 – 82
[17] O. Goksu, M. Altin, J. Fortmannand, P. Sorensen 2016 Field validation of IEC 61400-27-1 wind generator type 3 model with plant power factor controller *IEEE Trans on Energy Conv.* **31** 1170 –78
[18] P. C. Krause, O. Wasynczuk and S. D. Sudhoff 2002 *Analysis of Electric Machinery and Drive Systems* (New Jersey; IEEE/John Willey and Sons Incorporation) pp 100 –197
[19] J. Rocabert, A. Luna, F. Blaabjerg and P. Rodriguez 2012 Control of power converters in AC microgrids *IEEE Trans on Power Electro.* **27** 4734 – 49
[20] B. Wu, Y. Lang, N. Zargari and S. Kouro. 2011 *Power Conversion and Controls of Wind Energy Systems* (New Jersey; IEEE/John Willey and Sons Incorproation) pp 49 –100
[21] T. O. Ajewole, W. A. Oyekanmi, A. A. Babalola, M. O. Omoigui. 2017 RTDS modelling of hybrid-source autonomous electric microgrid *Int. J. Emerging Elect. Power Syst.* **18** doi: 10.1515/ijeeps-2016-0157
[22] V. Agarwal, R. K. Agarwal, P. Patidar and C. Patki. 2010 A novel scheme for rapid tracking of maximum power point in wind energy generating systems *IEEE Trans on Energy Conv.* **25** 228 –36
[23] L. Chih-Hong 2013 Recurrent modified elman neural network control of pm synchronous generator system using wind turbine emulator of pm synchronous servo motor drive *Int. J. Elect Power and Energy Systems* **52** 143 – 60.

IOP Conf. Series: Earth and Environmental Science **154** (2018) 012015 doi:10.1088/1755-1315/154/1/012015

Using particle swarm optimization to enhance PI controller performances for active and reactive power control in wind energy conversion systems

M Taleb[1], M Cherkaoui[2] and M Hbib[1]

[1] Ph.D. Student, Mohamed V University, EMI, Rabat, Morocco
[2] Professor, Mohamed V University, EMI, Rabat, Morocco

Abstract. Recently, renewable energy sources are impacting seriously power quality of the grids in term of frequency and voltage stability, due to their intermittence and less forecasting accuracy. Among these sources, wind energy conversion systems (WECS) received a great interest and especially the configuration with Doubly Fed Induction Generator. However, WECS strongly nonlinear, are making their control not easy by classical approaches such as a PI. In this paper, we continue deepen study of PI controller used in active and reactive power control of this kind of WECS. Particle Swarm Optimization (PSO) is suggested to improve its dynamic performances and its robustness against parameters variations. This work highlights the performances of PSO optimized PI control against classical PI tuned with poles compensation strategy. Simulations are carried out on MATLAB-SIMULINK software.

1. Introduction

In real life, many optimization problems are difficult to solve by exact optimization methods, due to properties, such as dimensionality, multimodality, epistasis (parameters interactions), and non-differentiability. Hence, approximate algorithms are an alternative approach for these problems. They try to find near-optimum solutions to hard optimization problems, mimicking intelligent processes of behaviours observed from nature, sociology, thinking, and other disciplines [1].
Among these approximate algorithms, Nature-inspired algorithms (NIA) [2], [3] are currently the most exciting research areas, and they are continuously demonstrating exceptional strength in solving complex real-life problems. The two most popular ones are Genetic Algorithm (GA) [4] and PSO [5] that have been found to be highly competitive for solving a wide variety of optimization problems. In spite of both of them are nature-inspired algorithms, GA is an evolutionary algorithm [6] inspired by genetic evolution process based on best-to-survive criteria, while PSO is a swarm intelligence algorithm [7] that simulates the social behaviour of swarms such as flocks of birds or schools of fish. PSO and GA can both obtain high quality solutions, yet the computational effort required by PSO is less than the corresponding effort required by GA. In addition, PSO is easy to implement and there are few parameters to adjust [8].
That is why we expected in our previous paper [9] to use PSO approach to remedy to PI controller low transient time performances and weak robustness.
PI controller has been used to control active and reactive power of the most popular wind energy conversion system based on Doubly Fed Induction Generator (DFIG) for which stator is connected directly to the grid while the rotor is connected to grid through double bi-directional converters separated by a DC-Link [9] and sized to allow flowing only 30% of power via the rotor winding as

illustrated in figure 1. Bulk and cost, as well as copper losses are then reduced accordingly in comparison with other topologies [10], [11].

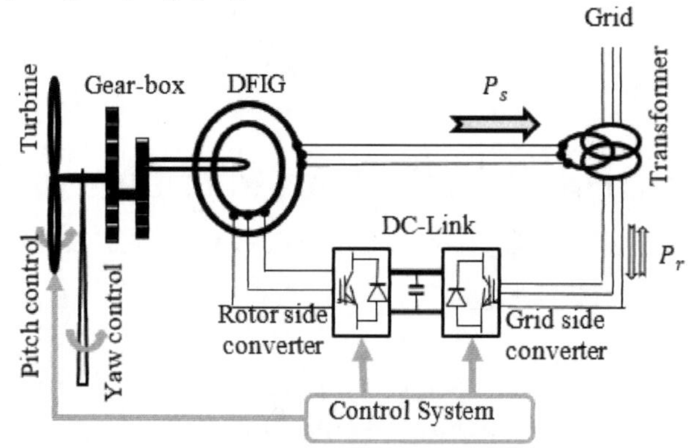

Figure 1. Synoptic Scheme of DFIG Wind Energy Conversion System.

This paper is organized in four major sections. In section 2, PSO paradigm is summarized and objective function to optimize is formulated. PSO optimized PI is presented in section 3. Finally, conclusions are given in last section (section 4).

2. PSO paradigm

PSO paradigm was first introduced by James Kennedy, a social psychologist, and Russell C. Eberhart, an electrical engineer, in 1995 [5]. Now, it is widely used in various fields because its easy implementation, low computational cost and good robustness [12], [13].

PSO evolves populations or swarms or individuals called particles which represents candidate solutions in search space. Each particle is assigned an initial random position vector and an initial random velocity. Then it updates iteratively its trajectory, namely position x_i and velocity v_i, according to its best experience p_i (personal best) and the best experience of the whole swarm g (global best).

Updating rule is very simple and requires only primitive mathematical operators as defined on following equations (1) and (2):

$$v_i(t+1) = wv_i(t) + c_1 r_1 (p_i(t) - x_i(t)) + c_2 r_2 (g(t) - x_i(t)) \tag{1}$$

$$x_i(t+1) = x_i(t) + v_i(t+1) \tag{2}$$

The three values that effect the new search direction, namely, current motion, particle own memory, and swarm influence, are incorporated via a summation approach as shown in equation (1) with three weight factors, namely, inertia factor (w), self-confidence factor (r_1c_1) and swarm confidence factor (r_2c_2), respectively. These latter two terms have randomized amplitudes due to random numbers r_1 and r_2 within $[0,1]$ that are generated at each iteration for each particle. Thus, different particle has different random numbers and then may have different strategies. Search space is then well explored and entrapment in local optima is avoided. Furthermore, to avoid big dispersion of the swarm that can lead to algorithm divergence, velocity magnitude should be controlled by setting a threshold $[v_{\min}, v_{\max}]$ on its absolute value. Finally, notion of "best" is evaluated, at each iteration, by the fitness function to be optimized (minimized or maximized). In present work, evaluation is made according to objective function given in equation (3).

$$ISE = \int \left[\left(P_{sref} - P_s\right)^2 + \left(Q_{sref} - Q_s\right)^2 \right] dt \tag{3}$$

ISE is the Integral Squared Error that can be replaced by ITAE (Integral Time Absolute Error) or IAE (Integral Absolute Error).

PSO tunes parameters of the controller to minimize the ISE expressed on both active and reactive powers basis. Other approaches can be investigated, depending in situations, such as minimizing ISE for active power only or reactive power only.
The three steps of velocity update, position update, and fitness calculations are repeated until a desired convergence criterion is met. In our case, algorithm stops when specified maximum iterations number is reached. The flowchart is given in figure 2.

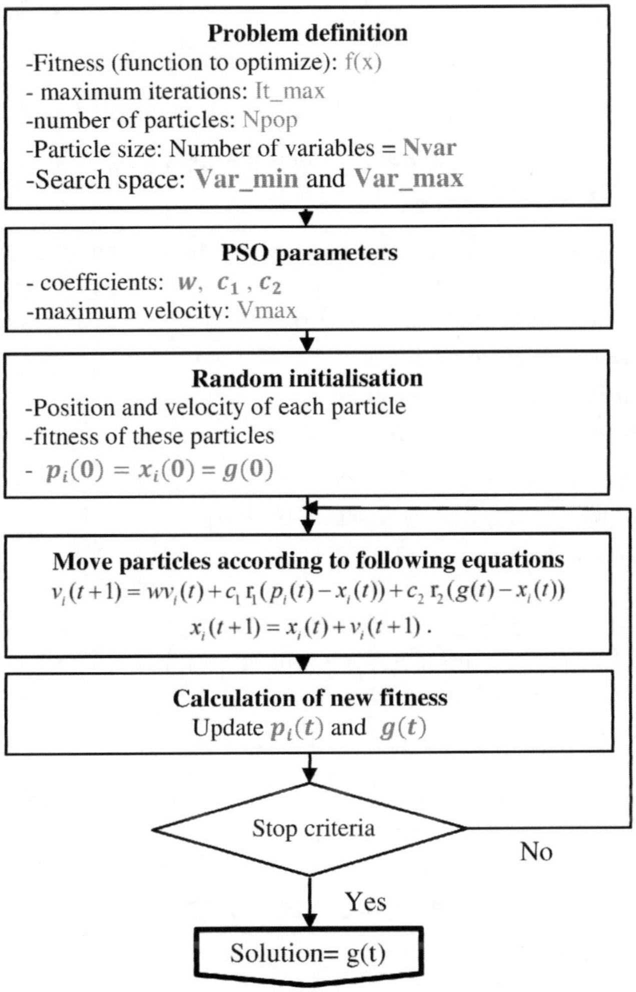

Figure 2. Flowchart of PSO algorithm.

3. PSO optimized PI controller

3.1. PI implementation

We have seen in [9] Direct and Indirect configurations of PI, and we highlighted the performances of indirect approach against the Direct one, mainly in term of overshooting rates. However, it was pointed out the poor capability of both configurations to deal with system uncertainties which may be caused by external disturbances and parameters variation.
We choose to continue deepen study for PI controller and we consider the Direct configuration only that we call back in figure 3.

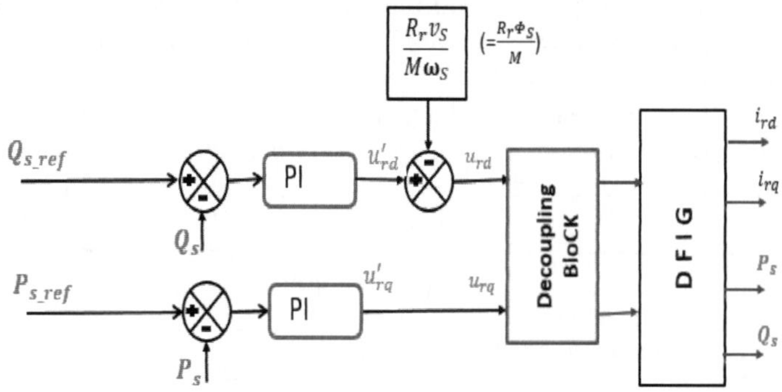

Figure 3. PQ Control using PI

3.2. PSO parameters

Compensation approach leads to same PI controller for both active and reactive powers. However, with PSO approach, we consider a general case where the two controllers can be totally different. Thus, four parameters will be considered: K_{pp} and K_{ip} the gains for active power PI control, K_{pq} and K_{iq} the gains for reactive power PI control.

Consequently, PSO algorithm is defined as follow:

- particles are the quadruples $\left(K_{pp}, K_{ip}, K_{pq}, K_{iq}\right)$,
- the swarm consists of 10 particles,
- each particle is randomly initialized in the interval $[0, 100]$,
- the cost function is defined by equation (3),
- the maximum number of iterations is 50.

Standard PSO is adopted for which parameters are $c_1 = c_2 = 2$ and $w = 1$. However, w is initialized to its standard value $w = 1$ and then gradually decreased (1% at each iteration) to avoid local optimum and stagnation problem due to fixed velocity step size [14].

3.3. Simulations and results

Even if in practice variations of P_{sref} and Q_{sref} are not instantaneous but made according to defined ramps, our simulations were made with profiles depicted in figure 4 where stiff variations have been imposed to the system. Results obtained in such hard conditions will be better in real case. P_{sref} is always negative (generator convention) while Q_{sref} can be positive (absorbed) or negative (delivered).

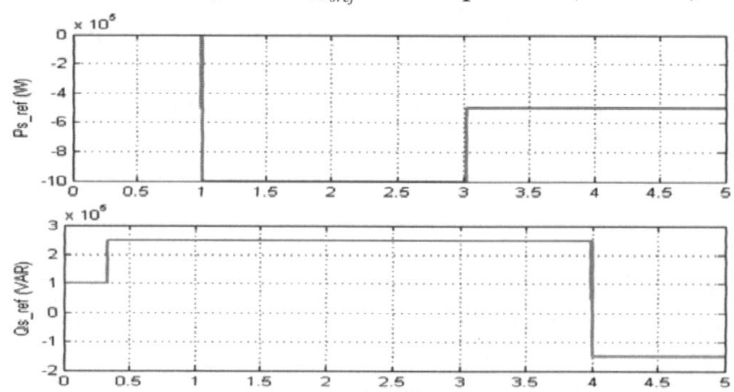

Figure 4. References P_{sref} and for Q_{sref} active and reactive power respectively.

Figure 5 and figure 6 depict the comparison results of classical PI (blue curve) and PSO optimized PI (red curve) for active and reactive power control respectively.

Optimization is achieved with $K_{pp} = K_{pq} = 0.1$, $K_{ip} = 0.0393$, and $K_{iq} = 3.10^{-7}$. ISE is reduced by 26% compared to classical PI (with poles compensation parameters).
Optimized PI provides good tracking without ripples that are due to residual coupling between active and reactive power. The high ripples observed during starting time (at less than 0.5s) for classical PI are also eliminated with PSO optimized PI.
The main parameters that influence the PI robustness are stator resistance and inductance. We already seen in [15] that robustness against R_s is quite acceptable but it noticed high oscillations amplitudes and a risk of dropping out when L_s vary by 10% to 20%. However, PSO Optimized PI maintains better robustness against same L_s variations as shown in figure 7.

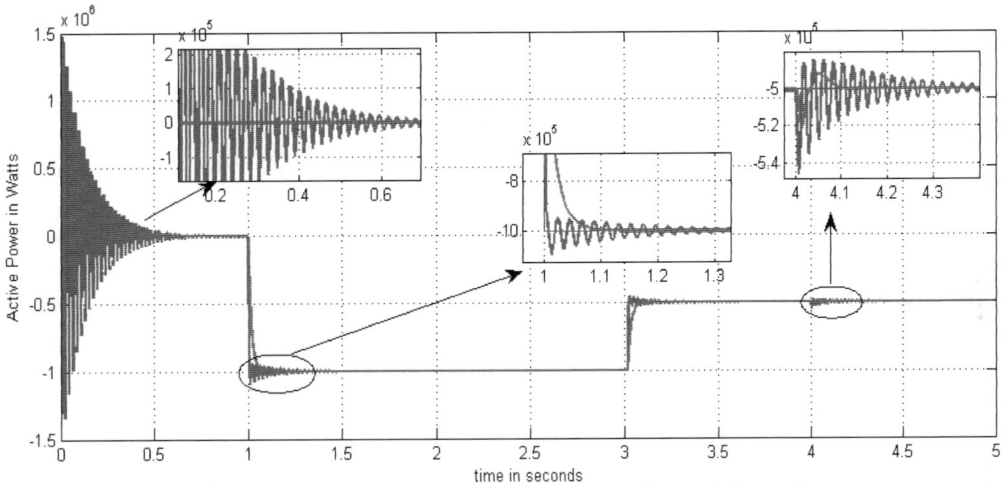

Figure 5. Comparison between classical PI and PSO optimized PI for active power control.

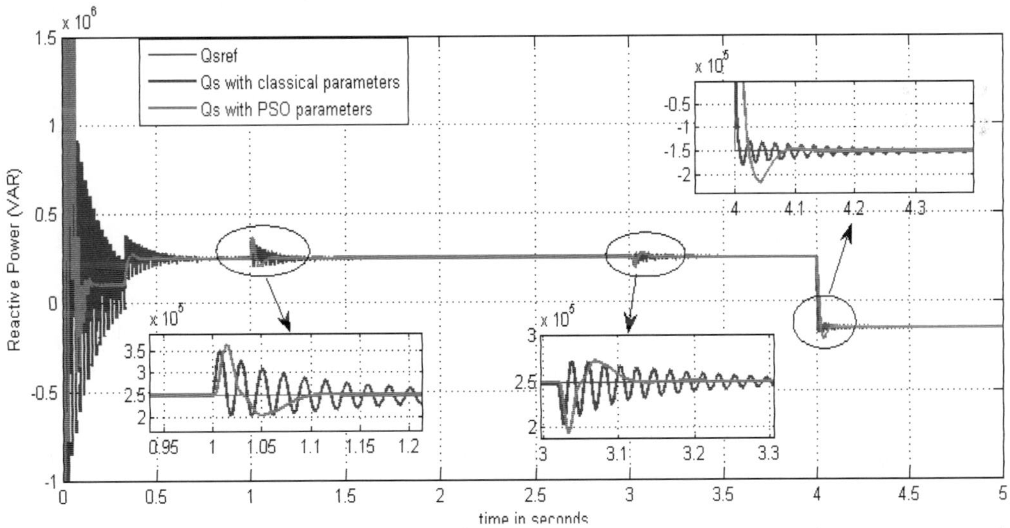

Figure 6. Comparison between classical PI and PSO optimized PI for reactive power control.

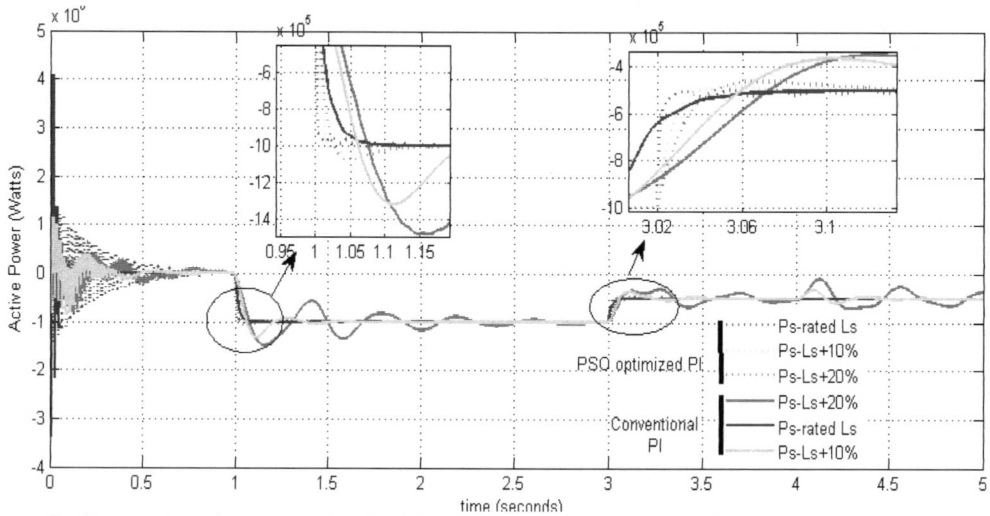

Figure 7. Comparison between classical PI and PSO optimized PI for robustness against stator inductance.

4. Conclusion

The main contribution of this work is to show how PSO can enhance PI performances for tracking any active and reactive powers profiles dictated by Grids Managers. High precision and fast response time are achieved by tuning PI gains that minimize the quadratic error (ISE). Robustness against parameters variations is also improved. This approach replaces the trial and error method that cannot succeed easily to satisfactory results.

PSO competes with many other nature-inspired algorithms and especially with GA in many fields. In addition, beside its easier implementation, it is generally considered efficient in terms of computational cost and robustness.

As prospect, this algorithm will be implemented in FPGA and DSP chips to appreciate its real-time response.

5. Appendices

DFIG parameters: $S_N = 3MVA$; $R_s = 0.012\Omega$; $R_r = 0.021\Omega$; $L_s = 0.0137H$ $L_r = 0.0136H$, $M = 0.0135H$, $p = 2$, $J = 0.07Kg.m?$; $f = 0.0024N.m.s^{-1}$. Turbine parameters: $P_{tN} = 4MW$ .Gear-box ratio: $G = 90$.

6. References

[1] K.-L. DU and M. Swamy, Search Optimization by Metaheuristics, Birkhaüser, July 2016.

[2] R. Ribeiro and F. Enembreck, "A sociologically inspired heuristic for optimization algorithms: A case study on ant systems," Expert Systems with Applications, vol. 40, pp. 1814-1826, 2013.

[3] R. A. Mahale and S. Chavan, "A Survey: Evolutionary and Swarm Based Bio-Inspired Optimization Algorithms," *International Journal of Scientific and Research Publications*, vol. 2, no. 12, 2012.

[4] J. H. Holland, "Genetic algorithms and the optional Allocation of trials," *SIAM Journal on Computing,* vol. 2, no. 2, pp. 88-105, 1973.

[5] R. Eberhart and J. Kennedy, "A New Optimizer Using Particle Swarm Theory," in Sixth *International Symposium on Micro Machine and Human Science*, IEEE, 1995.

[6] G. Rozenberg, T. Bäck and J. N. Kok, Handbook of Natural Computing, Springer-Verlag Berlin Heidelberg, 2012.

[7] S. Keerthi, K. Ashwini and M. Vijaykuma, "Survey Paper on Swarm Intelligence," *International Journal of Computer Application,* vol. 115, no. 5, pp. 8-12, April 2015.

[8] A. Layak and L. S. Samrat, "Particle swarm optimization with adaptive neighbourhood search for solving multi-objective optimization problems," *International Journal of Swarm Intelligence*, vol. 2, no. 1, 2016.

[9] M. TALEB and M. CHERKAOUI, "Active and Reactive Power Control of Doubly Fed Induction Generator Wind Turbines to Answer Grid Codes Requierements," *Journal of Clean Energy Technologies (JOCET)*, vol. 6, no. 2, pp. 101-105, 2018.

[10] O. P. Mahela and A. G. Shaik, "Comprehensive overview of grid interfaced wind energy generation systems," Renewable and Sustainable Energy Reviews, vol. 57, pp. 260-281, 2016.

[11] M. Cheng and Y. Zhu, "The state of the art of wind energy conversion systems and technologies: A review," Energy Conversion and Management, vol. 88, pp. 332-347, 2014.

[12] M. Taleb, A. Levant and F. Plestan, "Pneumatic actuator control: solution based on adaptative twisting and experimentation," Control Engineering Practice, 48 (5), 2013.

[13] J. Kennedy and Eberhart RC, "Particle Swarm Optimization," Proceedings of IEEE *International Conference on Neural Networks,* Perth, Australia, pp. 1942-1948, 1995.

[14] Y. Shi and R. Eberhart, "A Modified Particle Swarm Optimizer," in Proceeeeding of IEEE congress on evolutionary computation (CEC), AK,USA, 1998.

[15] M. TALEB and M. CHERKAOUI, "Active and Reactive Power Robust Control of Doubly Fed Induction Generator Wind Turbine to Satisfy New Grid Codes," in Advances in Intelligent Systems and Computing 565 (Springer), AECIA-2016, Marrakech, Morocco, 2018.

IOP Conf. Series: Earth and Environmental Science **154** (2018) 012016 doi:10.1088/1755-1315/154/1/012016

Maximum power point tracking techniques for wind energy systems using three levels boost converter

Cuong Hung Tran[1,2], Frédéric Nollet[1], Najib Essounbouli[1] and Abdelaziz Hamzaoui[1]

[1] CReSTIC, IUT de Troyes, 9 rue de Québec, 10000 Troyes, France
[2] E-mail : trancuonghung@gmail.com

Abstract. This paper presents modeling and simulation of three level Boost DC-DC converter in Wind Energy Conversion System (WECS). Three-level Boost converter has significant advantage compared to conventional Boost. A maximum power point tracking (MPPT) method for a variable speed wind turbine using permanent magnet synchronous generator (PMSG) is also presented. Simulation of three-level Boost converter topology with Perturb and Observe algorithm and Fuzzy Logic Control is implemented in MATLAB/SIMULINK. Results of this simulation show that the system with MPPT using fuzzy logic controller has better performance to the Perturb and Observe algorithm: fast response under changing conditions and small oscillation.

1. Introduction

Nowadays, among the all renewable energy sources, wind energy is one of the fastest expanding. It is a common source which not only for electrical power generation but also in rural sites using around the world. Small wind power systems are appropriate for remote areas.

The variable speed wind turbine technology using permanent synchronous generator (PMSG) is rapidly growing due to higher efficiency, lower mechanical stress and reduction in installation and maintenance costs [1]. Many power converter topologies are being developed for PMSG in wind energy conversion systems (WECS) to reduce cost, increase reliability and improve wind energy conversion efficiency [2].

In order to extract the maximum energy from the WECS, the generator rotational speed must be controlled by the MPPT algorithm; therefore, the MPPT is one of the factor importance in WECS. The goal of MPPT is to maximize the wind power capture at different wind speeds by adjusting the turbine speed. There are many different techniques for maximum power point tracking in the literature last decades [4]-[6], but two groups are most common. The first group requires prior knowledge of turbine parameters to calculate the operating point and the second group is based on iterative research.

The three level Boost converter (TLBC) have the advantages of the low stress, the low inductor current ripple and the low stitching loss as compared to conventional Boost converter [3]. Therefore, the TLBC are widely used in the modern power electronics applications, such as ac/dc PFC application [11], [12], dc/dc PV applications [3], the fuel-cell applications [13], [14] and the wind energy application [15].

2. Wind Turbine Systems

There are many possible configurations of power electronic converters and electrical generators for variable speed wind turbine systems. The figure 1 below shows the diagram of our wind energy conversion system.

- The wind turbine is directly coupled to the PMSG. Compared to other generators, the PMSG has advantage that it could couple directly to a wind turbine with no need for a gear box. WECS with PMSG can avoid the problem of wear and tear of gear, it can help wind turbine operate more reliable and reduce maintenance [7].
- The generated AC Voltage from PMSG cannot directly supply DC link so there are an uncontrolled rectifier and three level converter Boost DC-DC between the generator and DC bus.
- DC Resistive load

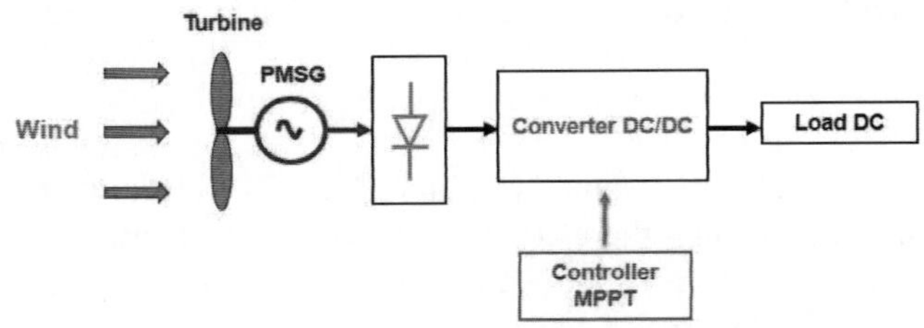

Figure 1. Configurations of Wind Power system

2.1. Model of wind turbine
Wind turbine is applied to convert the wind energy to mechanical torque. The conversion from wind speed to mechanical power (P_m) can be described in steady state by [8], [9]:

$$P_m = \frac{1}{2} C_p \rho \pi R_{blade}^2 v_{wind}^3 \tag{1}$$

In wind power equation (1), ρ is air density (1.225 kg/m^3), $R_{turbine}$ is the turbine radius in meters, v_{wind} is the wind speed in m/s, and C_p is the turbine performance coefficient.
In order to maximize the output power of a wind turbine, we have to optimize the value of C_p. The coefficient of performance is a strong and nonlinear function of the "Tip Speed Ratio". It depends on such factors as the number of blades, the pitch and shape of the blades. The maximum value C_p can attain around 59% theoretically. This coefficient is also known as Betz limit. The speed ration λ is defined:

$$\lambda = \frac{\Omega R_{turbine}}{v_{wind}} \tag{2}$$

Output torque of the turbine is calculated:

$$T_m = \frac{P_m}{\Omega} = \frac{1}{2} \frac{C_p(\lambda, \beta) \rho_{air} \pi R_{blade}^2 v_{wind}^3}{\Omega} \tag{3}$$

If the speed ratio λ is maintained at its optimal value λ_{opt}, the power coefficient is at its maximum value $C_{pM}=C_p(\lambda_{opt})$, the maximum power of the wind turbine will be:

$$P_m^{opt} = \frac{1}{2} C_{pM} \rho \pi R_{blade}^2 v_{wind}^3 \tag{4}$$

When the speed ratio assumed to be maintained at the optimum value, we obtain the optimum speed rotor:

$$\lambda^{opt} = \frac{\Omega R_{blade}}{v_{wind}} \Rightarrow \Omega^{opt} = \frac{\lambda^{opt} v_{wind}}{R_{blade}} \tag{5}$$

Thus, for each wind speed, there is a maximum rotor speed Ω_{opt} which made a maximum power recovered from the wind turbine as shown in the figure 2.

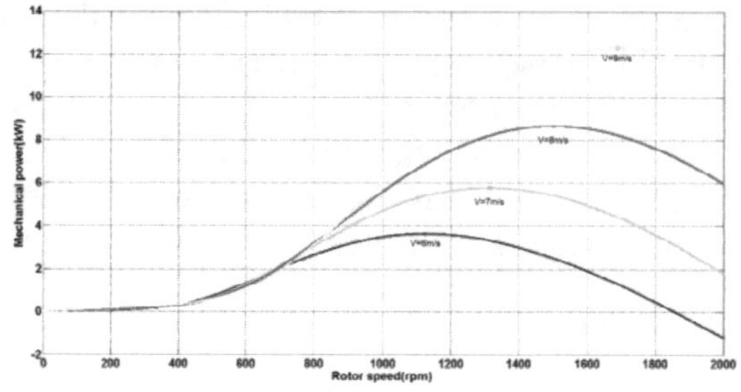

Figure 2. Power versus rotor speed at various wind speeds

2.2. Rectifier

In order to supply the DC-link, the PMSG should be connected to a rectifier that converts the AC voltage into DC voltage. In the literature, there are two possible configurations:

- Generator - diode rectifier - chopper - DC link
- Generator - PWM rectifier -DC link

In this study, we work with the first configuration.

2.3. Model of three-level boost converter

In the diagram of TLBC as shown in the figure 3, the input is a DC voltage source in series with an inductor L. The resistive load is connected across the capacitors to provide output voltages V_0. The midpoint of the capacitors C_1 and C_2 are connected to the centre point of the switches S_1 and S_2. By cause of the input inductor L and two diodes D_1 and D_2 in the TLBC, both switches can be turning ON at the same time without the concern of the sort-circuit damage. The ideal inductor and the ideal capacitors are assumed.

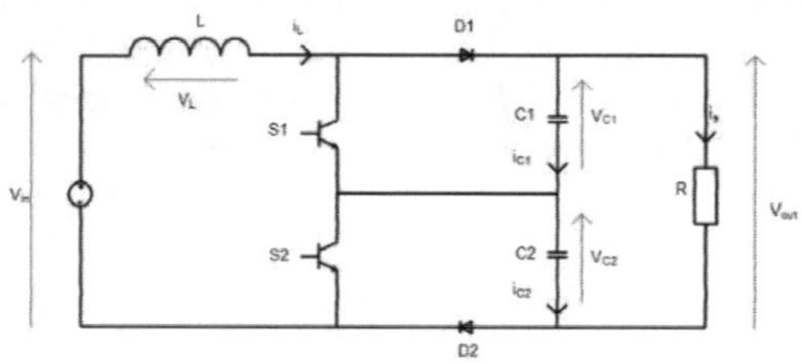

Figure 3. Model of three level Boost converter

2.4. Mode operation

The TLBC will be operated in four different modes according to the power switches conduction state, which can be both conducting, or both turned off, or one of them conducting and the other turned off as shown in the figure 4a, 4b, 4c, 4d.

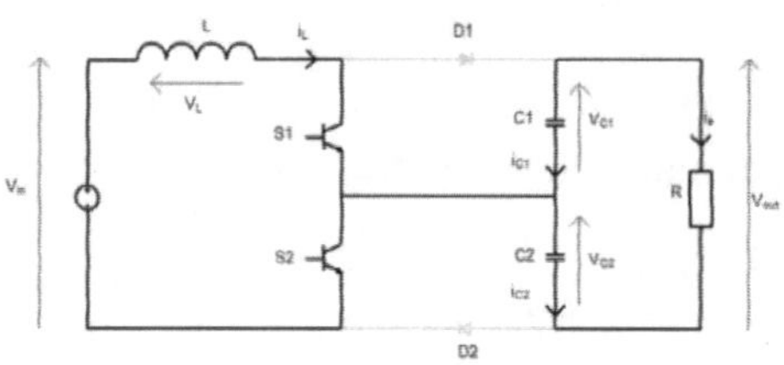

Figure 4a. S1-ON and S2-ON

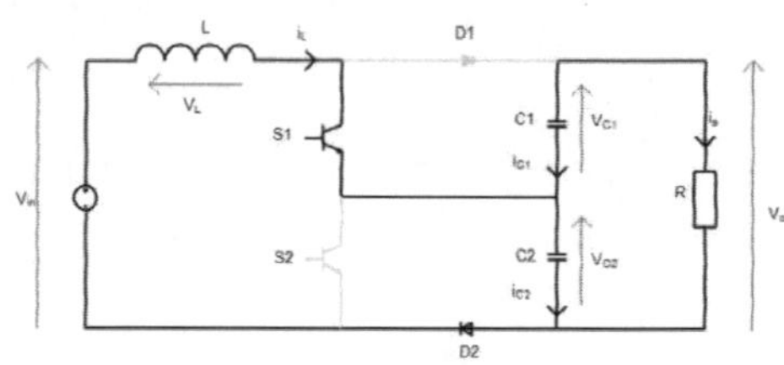

Figure 4b. S1-ON and S2-OFF

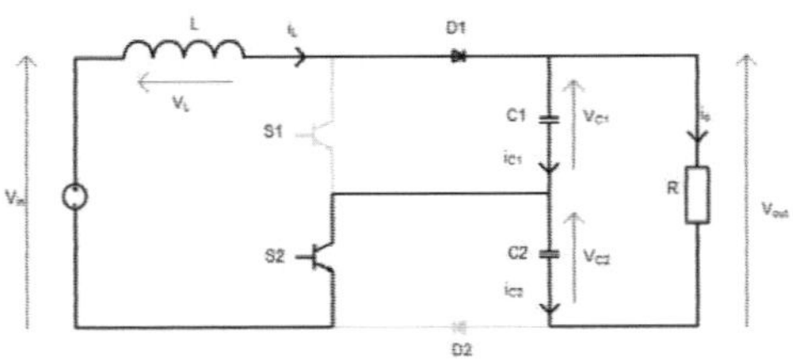

Figure 4c. S1-OFF and S2-ON

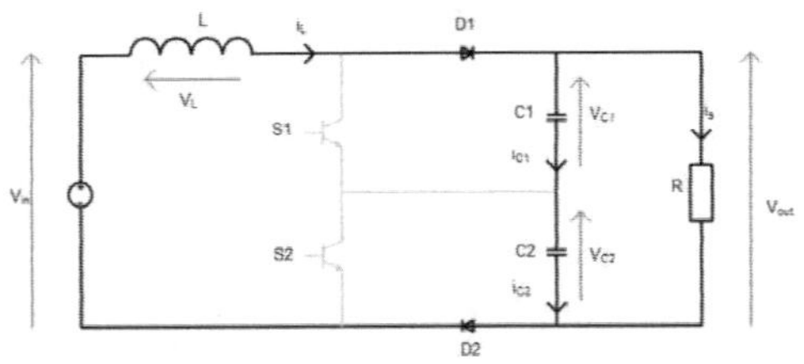

Figure 4d. S1-OFF and S2-OFF

2.5. Operation principles

The converter operated depends on the input voltage, which is lower or higher than half of the output voltage:

- <u>In region 1</u>: $V_{in} > V_{out}/2$: In both of mode 3 and 2, as $V_L > 0$, the inductor current raising polarity is positive. This occur only when duty ratio d_1 and d_2 are less than 0.5. In this region, both of two switches must not be "ON" at the same time; thus, only switching mode 2, mode 3 and mode 4 can be found.
- <u>In region 2</u>: $V_{in} < V_{out}/2$: In both of modes 3 and 2, as $V_L < 0$, the inductor current raising polarity is negative. Both of two switches must not be "OFF" at the same time; thus, only switching mode 1, mode 2 and mode 3 can be found.

In this study, we work in the symmetrical operation and continuous conduction mode by choosing $C_1 = C_2$ and the duty ratio $d_1 = d_2 = d$.

Equations to calculate a TLBC:

- The voltage gain of TLBC:

$$\frac{V_{out}}{V_{in}} = \frac{1}{1-d} \tag{6}$$

- Inductor current ripple:

$$\begin{cases} \Delta I_L = \dfrac{V_{in}(1-2d)d}{2Lf(1-d)} & for\ d < 0,5 \\ \Delta I_L = \dfrac{V_{in}(2d-1)}{2fL} & for\ d > 0,5 \end{cases} \tag{7}$$

- Output voltage ripple:

$$\begin{cases} \Delta V_{out} = \dfrac{V_{in}(1-2d)}{RCf(1-d)} & for\ d < 0,5 \\ \Delta V_{out} = \dfrac{V_{in}(2d-1)}{RCf(1-d)} & for\ d > 0,5 \end{cases} \tag{8}$$

3. Proposed MPPT methods

3.1. Perturbation and observation control

Perturbation and Observation (P&O) is the simplest MPPT algorithm that does not require any prior knowledge of the system, the turbine's characteristic curve or any additional sensor except the measurement of the power which is subjected to maximization. The principle of P&O algorithm is kept perturbing the control variable in the same direction until the power is decreased. The flowchart and principle of P&O method are shown in the figure 5 and 6 respectively.

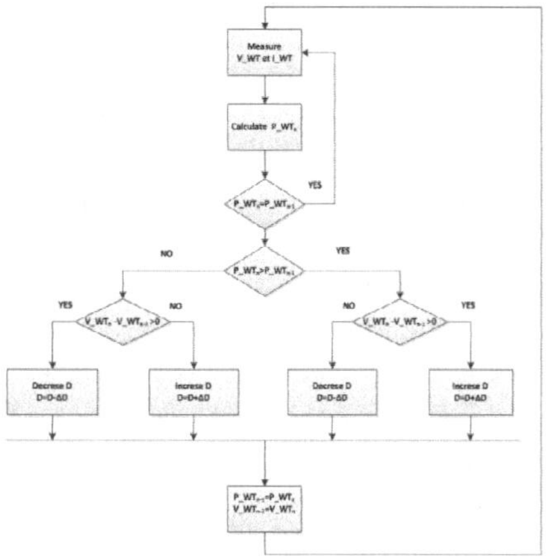

Figure 5. Flowchart of the P&O method

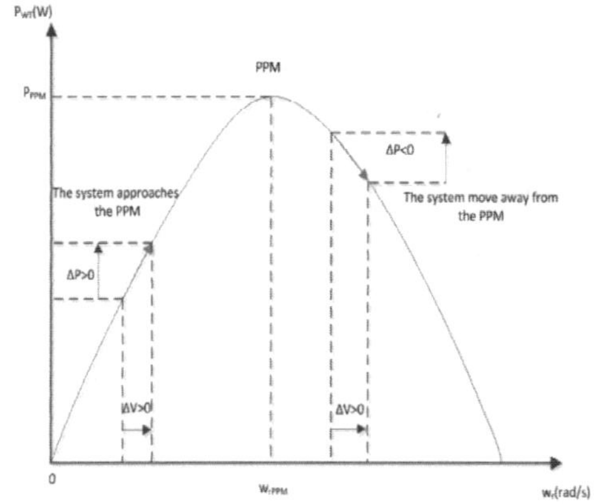

Figure 6. Principle of P&O method

In P&O method, choosing an appropriate step size is very important task. A larger step size leads to a faster response but more oscillations around the MPPT point. In reversed, a smaller step-size improves efficiency but reduces the convergence speed which can be incapable of tracking MPP under rapidly varying wind conditions.

3.2. Fuzzy control logic

The advantages of fuzzy logic controller (FLC) over the conventional methods are: (a) it does not need an accurate mathematical model; (b) it can work with imprecise inputs; (c) it can handle nonlinearity; and (d) it is more robust than conventional nonlinear controllers [16].

Our fuzzy logic controller is designed to vary the duty cycle of the converter to track the optimum rotor speed, thus maximizing the power recovered by the turbine. Our method based also on principle:

- On the MPP, we have $dP_{WT}/dV_{WT}=0$.
- If we are in the region $dP_{WT}/dV_{WT}>0$, we should increase the voltage or decrease the duty cycle to reach the MPP.
- If we are in the region $dP_{WT}/dV_{WT}<0$, we should decrease the voltage or increase the duty cycle to reach the MPP.

Fuzzy logic controller (FLC) MPPT method consists of four major elements: fuzzification, rules, interference engine and defuzzification [10].

Input and Output Variables: To implement the FLC for MPPT algorithm, the input and output variables of the FLC have to be determined. In this paper, we need two inputs: change in wind turbine power (dP/dV) and its derivative, output is duty cycle D of the three level Boost converter.

The output of the FLC is the DC/DC converter duty cycle given as:

$$D(n) = D(n-1) + \Delta D(n) \tag{9}$$

Membership Functions: In the FLC, the input and output variables are expressed by linguistic variables. The linguistic terms used here are:

- dP/dV [VeryNegative, Negative, Zero, Positive, VeryPositive]
- (dP/dV)' [Negative, Zero, Positive]

The five various terms of (dp/dV) and three terms of its derivative (dP/dV)' are shown in the figure 7 and 8 respectively.

The Control Rules: The control rules are derived from the experience or knowledge on the control system. The fuzzy rules are defined as follows:

R_i: IF (dP/dV) is A_i and (dP/dV)' is B_i, THEN $\Delta D(k+1)$ is C_i

Where A_i and B_i is the fuzzy subset, and C_i is a fuzzy singleton.

IOP Conf. Series: Earth and Environmental Science **154** (2018) 012016 doi:10.1088/1755-1315/154/1/012016

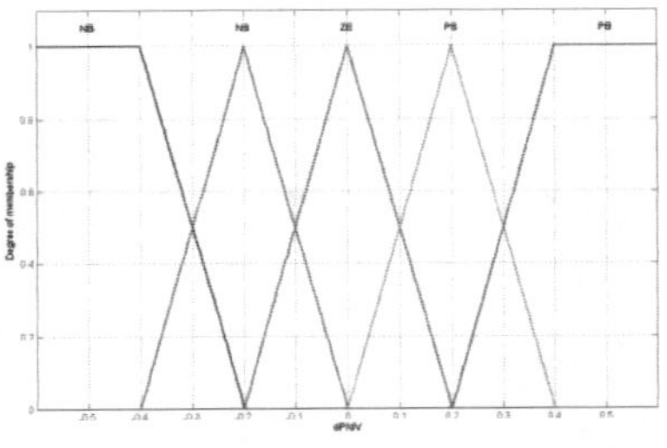

Figure 7. Membership Function (dP/dV)

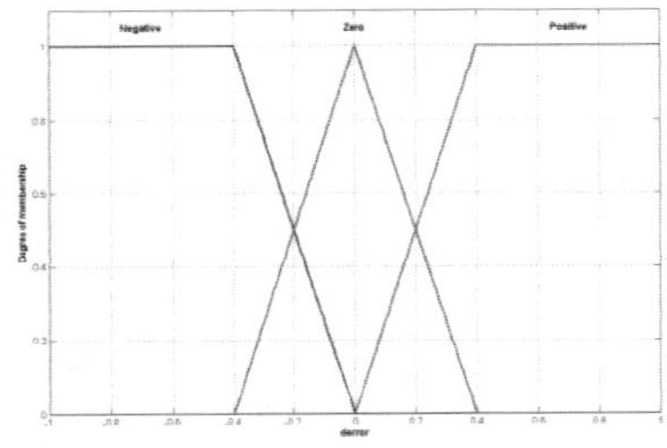

Figure 8. Membership Function (dP/dV)'

To get the output of inference, many methods are proposed by the researchers. There are some known methods such as Mandani, Sugeno, and Larsen. This paper used the min-max inference method and Takagi-Sugeno system.

Fuzzy rules are designed to achieve zero error at the state of the MPP. The main idea of the rule is to bring operating point of MPP by increasing or decreasing the duty ratio D depending on the position of the operating point from the MPP. If the operating point is distant from the MPP, the duty ratio will increase or decrease largely. In the proposed fuzzy controller, fifteen rules are formed, and it shown in the table 1:

Table 1. Rules of ΔD

ΔD		$(dP_{WT}/dV_{WT})'$		
		Negative	Zero	Positive
dP_{WT}/dV_{WT}	NB	3%	3%	3%
	NS	3%	1%	1%
	ZE	0%	0%	0%
	PS	-1%	-1%	-3%
	PB	-3%	-3%	-3%

Defuzzification: After the fuzzification, the defuzzification is performed which converts the fuzzied value into defuzzied value. It gives the final output value. In this paper, the center gravity defuzzification method is used.

The weighting factor is obtained by minimum operation, which is given by:

$$w_i = \min\left\{\mu_{dP/dV}, \mu_{(dP/dV)'}\right\} \tag{10}$$

The final output of the system is the weighted average of all rules output:

$$\Delta D(k) = \frac{\sum_{i=1}^{N} \omega_i C_i}{\sum_{i=1}^{N} w_i} \tag{11}$$

4. Simulation and results

4.1. Simulation

The system described in the section 2 is implemented Matlab Simulink as show figure 9.

IOP Conf. Series: Earth and Environmental Science **154** (2018) 012016 doi:10.1088/1755-1315/154/1/012016

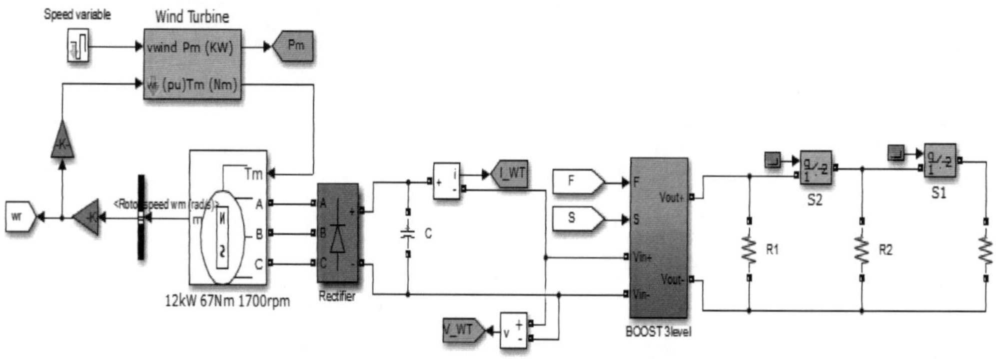

Figure 9. Implemented MATLAB Simulink

The model of wind turbine subsystem is built based on equation (1) - (4). Wind speed and load demand are varied within 80 seconds to test our controllers in various climatic and operating conditions. Figure 10 and figure 11 show our proposed fuzzy logic controller and P&O controller implemented in Simulink.

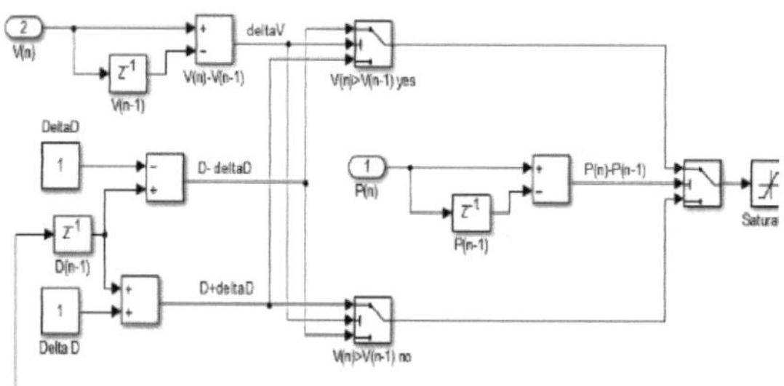

Figure 10. P&O Control

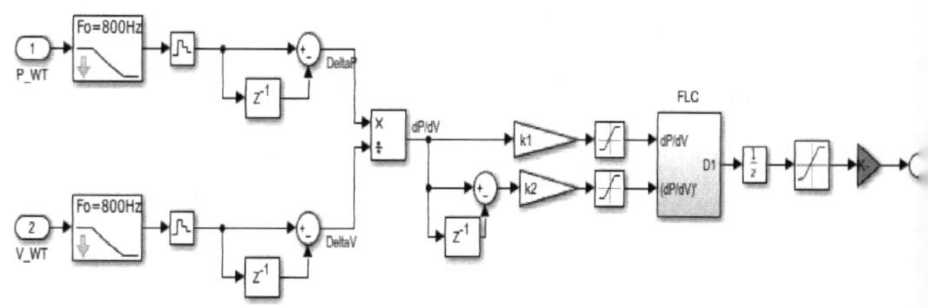

Figure 11. FLC Control

4.2. Results

In the first twenty seconds, the wind speed is 8m/s. According to the generator curve in the figure 2, we have the optimal rotor speed and the maximum power output is around 1500 rpm and 8.64 kW respectively. Our controller has chosen the good value of D in order to keep the rotor speed at around 1496 rpm, which make power attracted around 8.639 kW. From 20th seconds to 40th seconds, when wind speed change from 8 m/s to 6 m/s (for an optimal rotor speed of 1126 rpm and maximum power output of 3.64 kW), the wind turbine system moves toward to the new MPP. The generator speed changes to new optimal value as the figure 13. Our controller increases rotor speed at around 1117 rpm, which make power around 3.639 kW. Other tests are applied when wind speed changes from 6 m/s to 9 m/s and finally 9m/s to 7 m/s all the 20 seconds. We notice well that the rotor speed change according to the wind speed variation (at the 20th second, 40th second, 60th second) in order to have the maximum power output. On the other side, the figure 12 and 13 shows that our controller also works well to track the maximum power point when load demand change (30th and 70th second). These results still have some oscillation in P&O method because of the complex and nonlinear nature of the WECS, but it does not affect the final objective of our controller. Compared with conventional MPPT method, a FLC method has not only fast response under changing climatic or operating conditions but also small oscillation at the maximum power point.

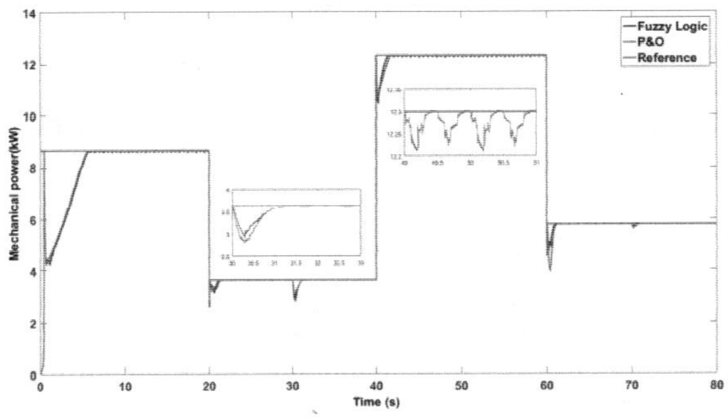

Figure 12. Mechanical power

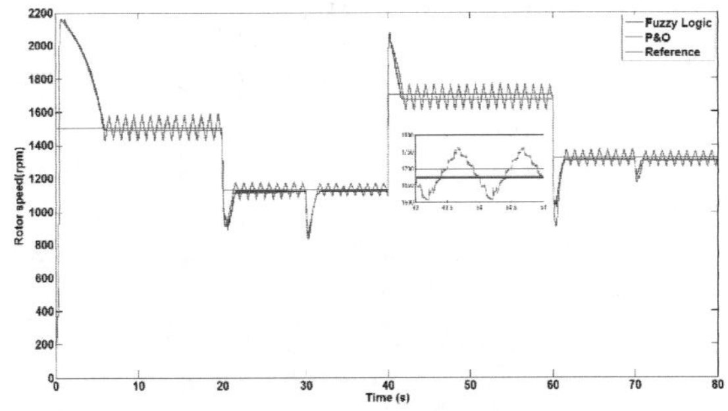

Figure 13. Rotor speed

5. Conclusion

This paper presents the modeling and simulation of three level Boost converter in Wind Energy Conversion System. Based on the simulation it can be concluded that with the both controllers can track the maximum power however the FLC has better performance to the P&O: fast response under changing conditions and small oscillation. In the future, we think about the case of failure of converter, for example an open-circuit fault. The TLBC will allow to the system go into degraded mode to provide a convenient service of nominal behavior during an open-circuit fault state by reconfiguring to the conventional DC-DC Boost converter.

6. References

[1] B. Wu, editor. 2011. *Power conversion and control of wind energy systems*. Piscataway, NJ : Hoboken, N.J: IEEE Press ; Wiley; 453 p. (IEEE Press series on power engineering).

[2] F. Blaabjerg, F. Iov, Z. Chen, K. Ma. 2010. *Power electronics and controls for wind turbine systems*. In: Energy Conference and Exhibition (EnergyCon), 2010 IEEE International; p. 333–344.

[3] C. Balakishan, N. Sandeep, M. V. Aware. 2015. *Design and Implementation of Three-Level DC-DC Converter with Golden Section Search Based MPPT for the Photovoltaic Applications*. Advances in Power Electronics:1–9.

[4] R. Kot, M. Rolak, M. Malinowski. 2013 May. *Comparison of maximum peak power tracking algorithms for a small wind turbine*. Mathematics and Computers in Simulation.91:29–40.

[5] M. A. Abdullah, A. H. M. Yatim, C. W. Tan, R. Saidur. 2012 Jun. *A review of maximum power point tracking algorithms for wind energy systems. Renewable and Sustainable Energy Reviews*.16(5):3220–7.

[6] D. Kumar, K. Chatterjee. 2016 Mar *A review of conventional and advanced MPPT algorithms for wind energy systems*. Renewable and Sustainable Energy Reviews.55:957–70.

[7] A. B. Cultura, Z. M. Salameh. 2011 *Modeling and simulation of a wind turbine-generator system*. In: Power and Energy Society General Meeting, IEEE; 2011. p. 1–7.

[8] E. Hau. 2013 *Wind Turbines*. Berlin, Heidelberg: Springer Berlin Heidelberg;

[9] M. Stiebler. 2008. *Wind energy systems for electric power generation*. Berlin: Springer; 193 p. (Green energy and technology).

[10] S. N. Sivanandam, S. Sumathi, S. N. Deepa 2007. *Introduction to fuzzy logic using MATLAB*. Berlin; New York: Springer; 430 p.

[11] J. R. Pinheiro, D. L. R. Vidor, H. Gründling, others. 1996. *Dual output three-level boost power factor correction converter with unbalanced loads*. In: Power Electronics Specialists Conference, 1996 PESC'96 Record, 27th Annual IEEE; p. 733–739.

[12] M. T. Zhang, Y. Jiang, F. C. Lee, M. M. Jovanovic. 1995. *Single-phase three-level boost power factor correction converter*. In: Applied Power Electronics Conference and Exposition, APEC'95 Conference Proceedings 1995, Tenth Annual IEEE. p. 434–439.

[13] B. Huang, A. Shahin, J. P. Martin, S. Pierfederici, B. Davat. 2008. *High voltage ratio non-isolated DC-DC converter for fuel cell power source applications.* In: Power Electronics Specialists Conference, PESC 2008 IEEE. p. 1277–1283.

[14] M. D. Bougrine, A. Benalia, M. H. Benbouzid 2015. *Simple sliding mode applied to the three-level boost converter for fuel cell applications*. In: Control, Engineering & Information Technology (CEIT), 2015 3rd International Conference on. IEEE; p. 1–6.

[15] V. Yaramasu, B. Wu. 2011. *Three-level boost converter based medium voltage megawatt PMSG wind energy conversion systems*. In: Energy Conversion Congress and Exposition (ECCE), IEEE. p. 561–567.

[16] V. S. C. Raviraj, P. C. Sen. 1997. *Comparative study of proportional-integral, sliding mode, and fuzzy logic controllers for power converters.* IEEE Transactions on Industry Applications; 33(2):518–524.

IOP Conf. Series: Earth and Environmental Science **154** (2018) 012017 doi:10.1088/1755-1315/154/1/012017

Environmental profile evaluations of piezoelectric polymers using life cycle assessment

M. A. Parvez Mahmud, Nazmul Huda, Shahjadi Hisan Farjana, and Candace Lang

Sustainable Energy Systems Engineering Group, School of Engineering, Macquarie University, Sydney, NSW 2109, Australia
E-mail: m-a-parvez.mahmud@mq.edu.au; nazmul.huda@mq.edu.au; shahjadi-hisan.farjana@mq.edu.au; candace.lang@mq.edu.au

Abstract—Piezoelectric materials are indispensable to produce electricity, harvesting ambient mechanical energy through motion for sectors and products, from sensors, to biomedical systems, to tiny electronics. Nylon 66 and tetrafluoroethylene dominate the market among thousands of piezoelectric materials to provide an autonomous power supply. Emphasis has been given on investigating the environmental impacts of both materials due to the growing consciousness of the ecological and health risks of piezoelectric polymers. The fabrication steps of these polymers from raw materials are extremely hazardous to the environment in terms of toxicity and human health effects. However, no quantification of the possible environmental impacts for the manufacturing of nylon 66 and tetrafluoroethylene exists. This research paper addresses their comparative environmental effects, in terms of chemical constituents. Life cycle impact analysis has been carried out by ReCipe 2016 Endpoint, Ecopoints 97, Raw material flows and CML-IA baseline methods, using Australasian life cycle inventory database and SimaPro software. The impacts are considered in categories like global warming, eutrophication, terrestrial ecotoxicity, human carcinogenic toxicity, fine particulates, and marine ecotoxicity. The results show that there is a significant environmental impact caused by tetrafluoroethylene in comparison with nylon 66 polymer during the manufacturing process. These impacts occur due to the quantity of toxic chemical elements present as constituents of tetrafluoroethylene raw material and its fabrication periods. It can be anticipated that a better ecological performance can be attained through optimization, especially by cautiously picking substitute materials and machines, taking into account the toxicity aspects, and by minimizing the impacts related to designs, fabrication processes and usage.

1. Introduction

In recent times, there has been an increased demand for piezoelectric polymers to energize portable electronics, wearable sensors, biomedical tiny devices, and self-powered systems [1]. Nylon 66 and tetrafluoroethylene are better choices than lead-based piezoelectric materials for their smaller environmental impacts and health hazards [2]-[4]. The advancement of nature-friendly, viable piezoelectric polymer fabrication system is getting considerable interest because of its extensive usage and greenhouse gas (GHG) release reduction. However, taking the remarkable part that nylon 66 and tetrafluoroethylene material has to perform in fulfilling present power crisis, investigators need to assess their ecological impacts counting all the steps of the production method. The life-cycle assessment (LCA) is an orderly scientific way that provides an evaluation of the total ecological effects of a system for its whole lifetime using the comprehensive life-cycle input and output factors that act within the particular life-cycle inventory (LCI) [5]-[8].

All the piezoelectric polymer related LCA studies until now has stressed the assessment of the life-cycle GHG releases. None of the previous LCA research [9]-[10] assessed and compared the net environmental impacts by nylon 66 and tetrafluoroethylene polymer production technologies.

Moreover, no noteworthy study is stated which comprehensively highlights the ecological influences happened by the whole life cycle of these technologies. This work try to minimize this research gap by accomplishing a complete environmental LCA of nylon 66 and tetrafluoroethylene fabrication technologies. This research therefore is a first and inclusive life-cycle environmental impact assessment of nylon 66 and tetrafluoroethylene polymers through systematic life-cycle assessment approach considering their input materials, fabrication steps and associated machines. The LCI dataset is gathered from an Australasian life-cycle inventory (AusLCI) database and analysis is carried out using SimaPro software. LCA has been done by ReCipe 2016 Endpoint, Ecopoints 97, Raw material flows and CML-IA baseline methods. The impacts are addressed under 19 effect-assessing sets like global warming, acidification, eutrophication, terrestrial ecotoxicity, human toxicity, human carcinogenic toxicity, fine particulate matter formation, and marine ecotoxicity etc. The results indicate that nylon 66 has a better performance for most of the considered categories than tetrafluoroethylene. Moreover, the outcome by the Ecopoints 97 method shows that the emission rate of greenhouse gases (GHG) and CO2 gas is much higher in tetrafluoroethylene than for nylon 66. In Section 2, an overview of the LCA methods is highlighted. Section 3 reveals the LCA outcomes using the ReCipe 2016 Endpoint, Ecopoints 97, Raw material flows and CML-IA baseline methods, thus advises environmental-impact diminution approaches for considered technologies.

2. Methodology

There are a few key steps involved in fabricating the nylon 66 and tetrafluoroethylene polymers at a plant. Nylon 66 polymer membrane is manufactured by polycondensation of adipic acid and hexamethylene-diamine. The key step for nylon 66 formation are the ammonium dinitramide (ADN) process, the Hexamethylene diamine (HMD) process and the adipic acid process which is highlighted in Figure 1. On the other hand, tetrafluoroethylene (TFE) is a chemical compound of fluorine and carbon. The major stages for TFE polymer production are casting, drying and salt crystalizing, sintering, cooling, and salt leaching (Figure 1). LCA is an effective approach to methodically calculate the environmental effects throughout the entire procedure of a system [11]-[12].

LCA is done by collection of the inputs and outputs at every production steps of the material, and environmental impacts are calculated using four main LCA steps [13] namely: (i) goal and scope definition, where the aim is described and the LCA boundaries are fixed; (ii) Life cycle inventory (LCI) analysis where the inputs and outputs at every steps of the manufacturing processes are assembled; (iii) life cycle impact assessment, where output emissions and input resources are clustered into their particular impact groups and transformed into same units for comparative assessment; (iv) the interpretation of the LCI and effect evaluation outcomes to realize the aims of this research.

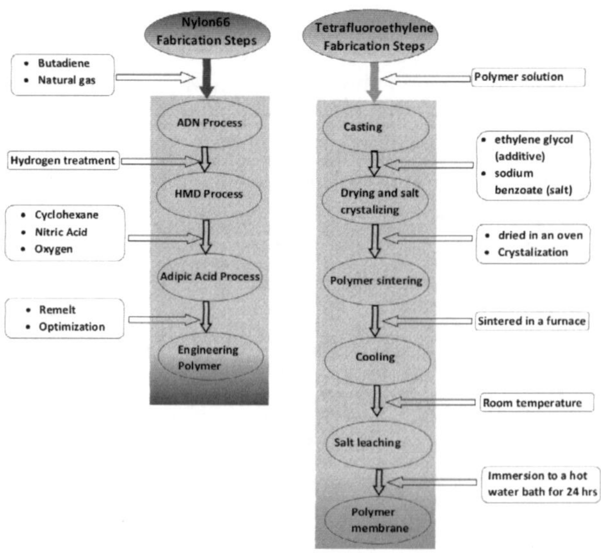

Figure 1. Nylon 66 and tetrafluoroethylene fabrication methods in plant.

To check the outcomes of every stage of both polymer production processes several sustainability metrics is considered into the inventory, and then LCA in this work is carried out by ReCipe 2016 Endpoint, Ecopoints 97, Raw material flows and CML-IA baseline methods, using an Australasian life-cycle inventory database and SimaPro software. GHG emissions, abiotic depletion, acidification, eutrophication, marine aquatic ecotoxicity, human toxicity, and ionization radiation rates are obtained and compared for both cases based on the CML-IA baseline method [14]. Several metal and gas based endpoint indicators (Cd, Hg, Zn, Pb, Cu, Cr, P, N, NOx, SOx, NH3, CO2, Dust etc.) are calculated and compared utilizing Ecopoints 97 method. The total inputs from nature and outputs to air, water and land of both processes are also measured and compared by the Raw material flows method. Finally, the ReCipe 2016 Endpoint method is used to get and compare 19 different environmental impact categories such as stratospheric ozone depletion, water consumption (human health, terrestrial ecosystem, acquatic ecosystems), global warming (human health, terrestrial ecosystem, acquatic ecosystems), fine particulate matter formation etc. for both cases.

3. Results and discussion

The effects are depicted at the effect calculation level. In the act of a comparative study, the results for each impact type for the system with the highest values are computed as 100% and the outcome for another arrangement is computed as the respective proportion.

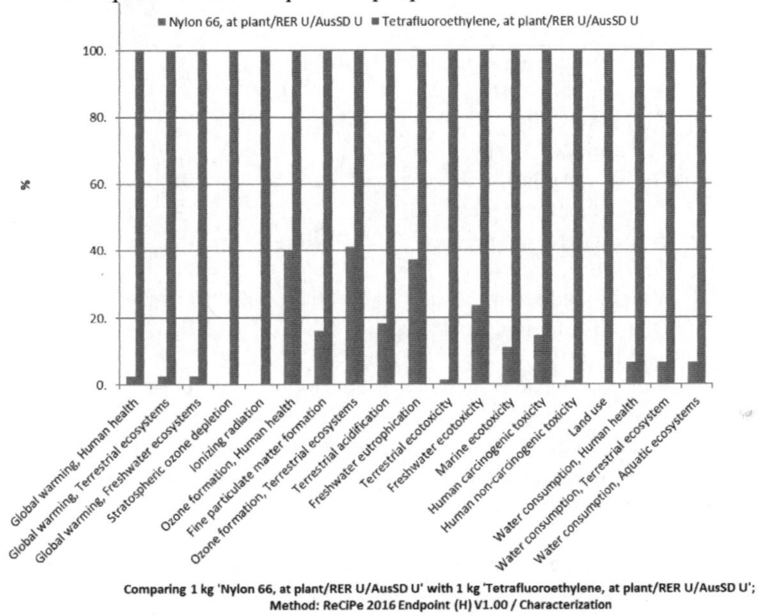

Figure 2. Impact comparison for 19 different impact categories of nylon 66 and tetrafluoroethylene fabrication processes in plant, by ReCiPe 2016 Endpoint method.

Figures 2 and 3 show the relative ecological impacts happened for each damage category of the nylon 66 and tetrafluoroethylene fabrication processes in plant. Figure 2 reveals the evaluation of net impacts, with the maximum effect set to 100, employing the ReCiPe 2016 Endpoint approach. It is evident that factories of the nylon 66 production process perform better environmentally than TFE fabrication process based plants. The endpoint comparison after weighing, using CML-IA baseline methodology is represented at Figure 3. The results indicate that the maximum impact happens at the eutrophication impact category, whilst the minimum is obtained in the abiotic depletion category and the marine aquatic ecotoxicity category by nylon 66 production process factories. It obtained because of the amount of toxic chemical elements available as constituents of TFE raw material and its fabrication stages.

The major impact categories such as global warming, ozone formation, acidification, eutrophication, ecotoxicity, human toxicity, water consumption, stratospheric ozone depletion, ionizing radiation, land

IOP Conf. Series: Earth and Environmental Science **154** (2018) 012017 doi:10.1088/1755-1315/154/1/012017

use, etc., which are found by ReCiPe 2016 Endpoint method and results are showed at Table 1. The overall results prove that greater environmental impact is caused by the TFE membrane polymer fabrication plant over the nylon 66 plant. Figure 4 shows the comparative metal and gas-based LCA outputs of nylon 66 and tetrafluoroethylene fabrication process in plant using Ecopoints 97 method. Among several categories CO2, NOx and SOx emission rates are higher than others such as P, Hg, Cd, Cr, Zn, N, Pb etc. The comparative LCA inputs and outputs of the nylon 66 and tetrafluoroethylene fabrication processes in plant using the Raw material flows methodology is highlighted in Figure 5. Nylon 66 plant takes 95% smaller inputs from nature than TFE plant outputs to air is highest (39%). Overall, the effects obtained from this work highlight that a large amount of hazardous emission is occurred from both the nylon 66 and tetrafluoroethylene fabrication processes in plant, which can be reduced by optimization of raw materials and used devices at different fabrication steps. Thus emission of harmful radioactive particles and gases to the environment and ecosystems can be reduced. LCA results-ReCiPe 2016 endpoint method

TABLE 1: LCA RESULTS- RECIPE 2016 ENDPOINT METHOD

Impact Category	Nylon 66, at plant/RER U/AusSD U	Tetrafluoroethylene, at plant/RER U/AusSD U
Global warming, Human health	2.4061	100
Global warming, Terrestrial ecosystems	2.405	100
Global warming, Freshwater ecosystems	2.4054	100
Stratospheric ozone depletion	0.0959	100
Ionizing radiation	0.0062	100
Ozone formation, Human health	40.3002	100
Fine particulate matter formation	16.0143	100
Ozone formation, Terrestrial ecosystems	41.0914	100
Terrestrial acidification	18.357	100
Freshwater eutrophication	37.2802	100
Terrestrial ecotoxicity	1.3289	100
Freshwater ecotoxicity	23.5896	100
Marine ecotoxicity	11.1644	100
Human carcinogenic toxicity	14.692	100
Human non-carcinogenic toxicity	0.9986	100
Land use	0.1047	100
Water consumption, Human health	6.478	100
Water consumption, Terrestrial ecosystem	6.478	100
Water consumption, Aquatic ecosystems	6.478	100

The assessment outcome of human toxicity and ecotoxicity of the nylon 66 and tetrafluoroethylene fabrication processes in plant is highlighted in Figure 6, which is obtained by USEtox method. The cancer based human toxicity is doubled in TFE membrane polymer fabrication plant with respect to Nylon 66 manufacturing plant.

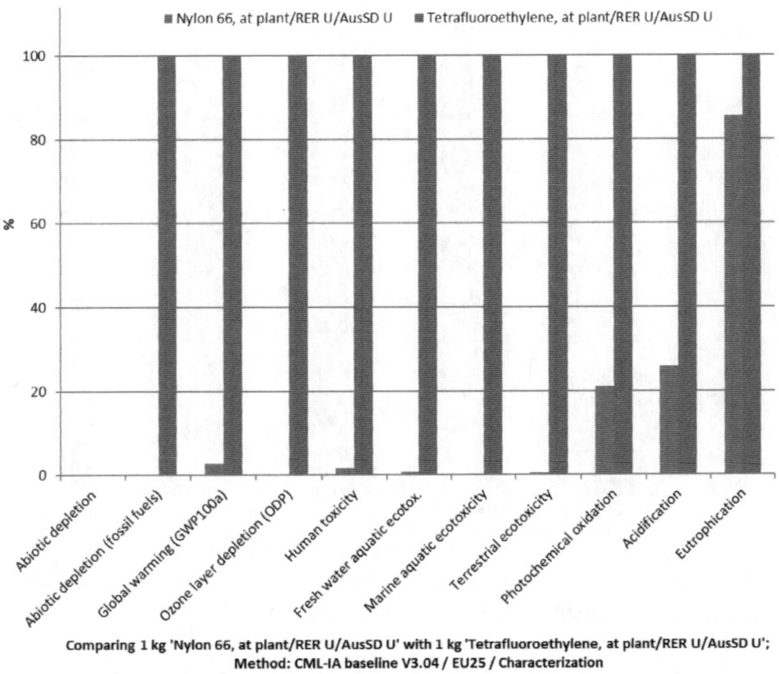

Figure 3. Comparative life-cycle impact assessment of the nylon 66 and tetrafluoroethylene fabrication processes in plant using the CML-IA baseline method.

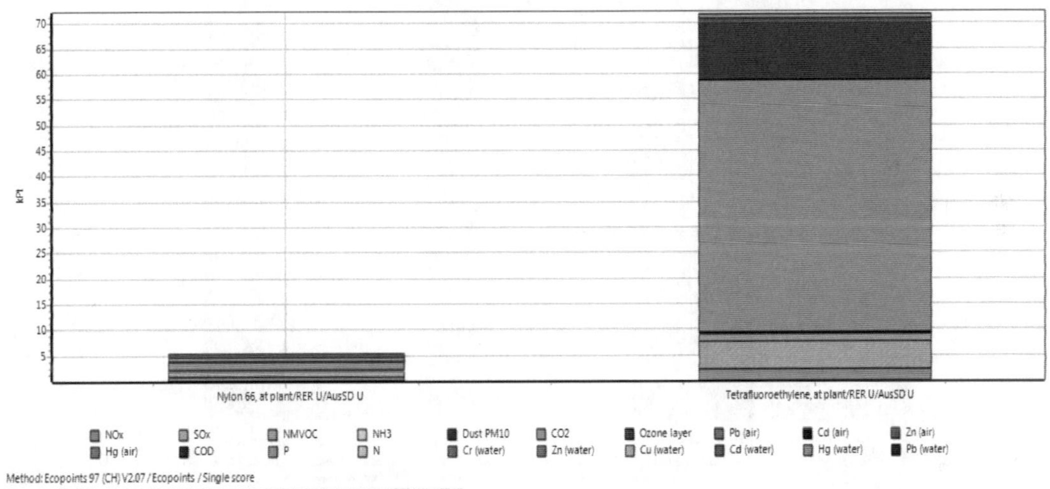

Figure 4. Comparative metal and gas-based LCA outputs of the nylon 66 and tetrafluoroethylene fabrication processes in plant using the Ecopoints 97 method.

Result indicates that the eco-toxicity occurred by Nylon 66 plant is one third of the amount in tetrafluoroethylene plant. It means that the production process of TFE polymer in plant is associated with more toxic element emissions compared to Nylon 66.

The relative effects of both polymer production plants on human health, ecosystem quality and resources, by Eco-indicator 99 method is represented at Figure 7. As shown, the TFE membrane polymer fabrication plant results in more damage compared to the nylon 66 plant in all three impact categories. This means that despite the potential application of tetrafluoroethylene there is still need for further improvement of its environmental profile through optimization during producing at plant.

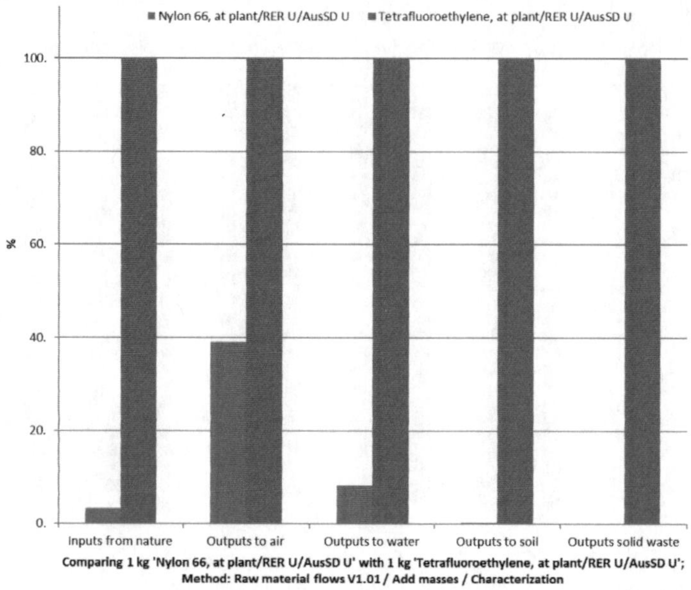

Figure 5. Comparative LCA inputs and outputs of the nylon 66 and tetrafluoroethylene fabrication processes in plant using the Raw material flows method.

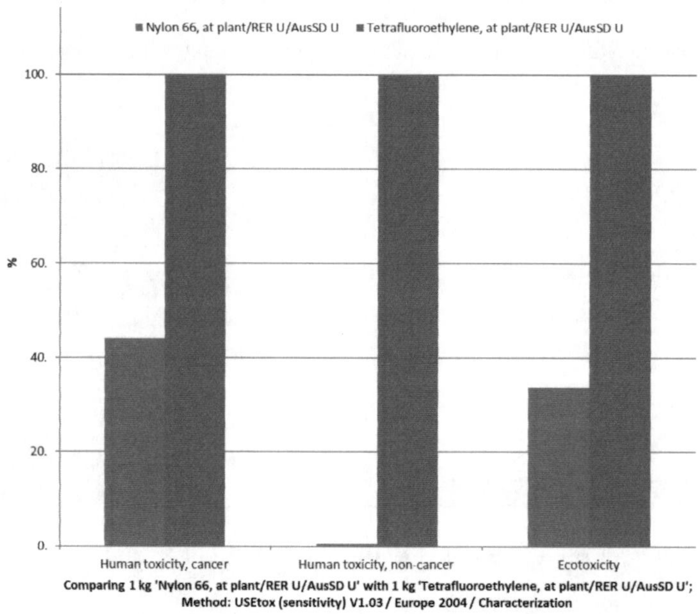

Figure 6. Comparative human toxicity and ecotoxicity of the nylon 66 and tetrafluoroethylene fabrication processes in plant using the USEtox method.

The comparative greenhouse gas emissions of the nylon 66 and tetrafluoroethylene fabrication processes in plant is depicted in Figure 8, which is obtained by the Greenhouse gas protocol methodology. We can conclude from this figure that a 96% reduction in GHG emissions is possible with Nylon 66 technologies and that plants score significantly better than the TFE membrane polymer production plants.

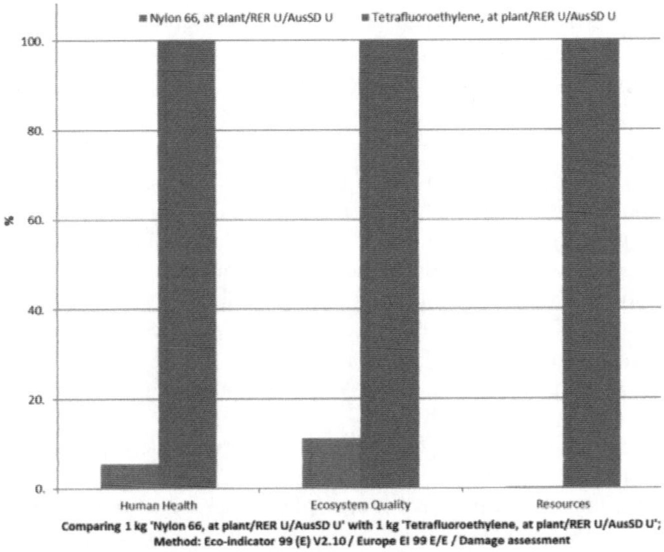

Figure 7. Relative effects of both polymer plants during production on human health, ecosystem quality and resources, by Eco-indicator 99 method.

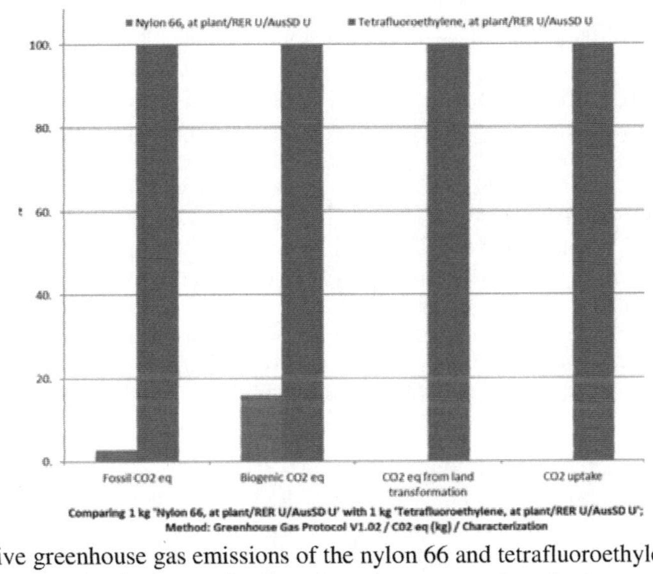

Figure 8. Comparative greenhouse gas emissions of the nylon 66 and tetrafluoroethylene fabrication processes in plant using the Greenhouse gas protocol method.

4. Conclusion

Piezoelectric materials like nylon 66 and tetrafluoroethylene are in great demand for self-powered electronics, biomedical devices and sensor production. The best choice of material depends on higher efficiency, lower fabrication cost, greater lifetime and better environmental performance. This research paper highlights the comparative environmental impacts of nylon 66 and tetrafluoroethylene through systematic LCA analysis by the ReCipe 2016 Endpoint, Ecopoints 97, Eco-indicator 99 (E), USEtox, Raw material flows and CML-IA baseline methods, using an Australasian life-cycle inventory database and SimaPro software. The results indicate that nylon 66 polymer fabrication plants emit lower GHGs, CO_2, NOx and SOx compared to TFE manufacturing plants. Moreover, global warming, ozone formation, acidification, eutrophication, ecotoxicity, human toxicity, water consumption, stratospheric ozone depletion and ionizing radiation rates are higher in tetrafluoroethylene polymer plants than for nylon 66. These impacts are occurred due to the amount of toxic chemical elements exist as constituents of tetrafluoroethylene raw material and its manufacturing stages. Therefore, nylon 66 is indeed a better choice of piezoelectric material than TFE. It is suggested that cautious choice of machines and required ingredients can reduce toxic emissions and increase environmental performance by polymer production plants.

5. References

[1] G. T. Hwang, M. Byun, C. K. Jeong, and K. J. Lee, "Flexible piezoelectric thin-film energy harvesters and nanosensors for biomedical applications," Adv. Healthcare Mater., vol. 4, pp. 646–658, April 2015.

[2] J. Koruza, B. Rozic, G. Cordoyiannis, B. Malic, and Z. Kutnjak, "Large electrocaloric effect in lead-free $K_{0.5}Na_{0.5}NbO_3$-$SrTiO_3$ ceramics," Appl. Phys. Lett., vol. 106, pp. 202905-1-3, May 2015.

[3] T. Lusiola, F. Bortolani, Q. Zhang, and R. Dorey, "Molten hydroxide synthesis as an alternative to molten salt synthesis for producing $K_{0.5}Na_{0.5}NbO_3$ lead free ceramics," J. Mater. Sci., vol. 47, pp. 1938-1942, July 2012.

[4] C. He, X. Li, Z. Wang, Y. Liu, D. Shen, T. Li, X. Long, and Z.-G. Ye, "Growth of $Pb(Fe_{1/2}Nb_{1/2})O_3$–$Pb(Yb_{1/2}Nb_{1/2})O_3$–$PbTiO_3$ piezo-/ferroelectric crystals for high power and high temperature applications," Cryst. Eng. Comm., vol. 14, pp. 4407–4413, May 2012.

[5] L. Gagnon and J.F. Vate, "Greenhouse gas emissions from hydropower: The state of research in 1996," Energy Policy, vol. 25, pp. 7-13, March 1997.

[6] P. Denholm and G.L. Kulcinski, "Life cycle energy requirements and greenhouse gas emissions from large scale energy storage systems," Energy Conversion and Management, vol. 45, pp. 2153-2172, August 2004.

[7] H. Hondo, "Life cycle GHG emission analysis of power generation systems: Japanese case" Energy, vol. 30, pp. 2042–2056, April 2005.

[8] M. Pehnt, "Dynamic life cycle assessment (LCA) of renewable energy technologies," Renew. Energy, vol.31, pp. 55-71, May 2006.

[9] A. Y. Hoekstra and T. O. Wiedmann, "Humanity's unsustainable environmental footprint," Science, vol. 344, pp. 1114-1117, June 2014.

[10] T. O. Wiedmann, S. Suh, K. Feng, M. Lenzen, A. Acquaye, K. Scott, and J. R. Barrett, "Application of hybrid life cycle approaches to emerging energy technologies--the case of wind power in the UK," Environ. Sci. Technol., vol. 45, pp. 5900–5907, July 2011.

[11] S. L. Koh, A. Genovese, A. A. Acquaye, P. Barratt, N. Rana, J. Kuylenstierna, and D. Gibbs, "Decarbonising roduct supply chains: design and development of an integrated evidence-based decision support system – the supply chain environmental analysis tool (SCEnAT)," Int. J. Prod. Res., vol. 51, pp. 2092–2109, June 2013.

[12] S. Hellweg and L. M. Canals, "Emerging approaches, challenges and opportunities in life cycle assessment," Science, vol. 344, pp. 1109–1113, June 2014.

[13] M. Finkbeiner, A. Inaba, R. Tan, K. Christiansen, and H.-J. Kluppel, "The new international standards for life cycle assessment: ISO 14040 and ISO 14044," Int J LCA, vol. 11, pp. 80–85, March 2006.

[14] Institute of Environmental Sciences, "CML-IA Methodology" 2015.

IOP Conf. Series: Earth and Environmental Science **154** (2018) 012018 doi:10.1088/1755-1315/154/1/012018

Application study of Bio-FGD based on environmental safety during the coal combustion

Pin Zhang

Institute of Resource and Safety Engineering, China University of Mining and Technology (Beijing), 100083, Beijing, China

Abstract. Coal combustion produces a large amount of acidic gas, which is the main cause of acid rain and other natural disasters. Flue Gas Desulfurization (FGD) is a necessary requirement for clean coal combustion. Compared with the traditional chemical desulfurization technology, biological desulfurization has the advantages of low operating cost, without secondary pollution, low carbon emission and the additional economic benefits. The process and structure of BioDeSOx which as one of Bio-FGD technology is introduced. The major factors that influent BioDeSOx Bio- FGD system is the pH, oxidation reduction potential (-300 MV to -400MV), electrical conductivity, the adding amount of nutrient and temperature (30ºC-40ºC). Taking the Bio- FGD project of Yixing xielian thermal power plant as an example, the BioDeSOx technology was applied in this project. The environmental and economic benefits of the project were greater than the traditional desulfurization technology. With the continuous improvement of environmental safety standards, Bio- FGD technology will have broad application prospects.

1. Introduction

Since 2000, China's rapid economic development has brought a growing demand for energy. Coal mining technology made considerable progress in the past 15 years by the exploitation of shallow and low sulfur coal seam turn to deep and high sulfur coal seam, especially in the southern areas of China where there is a coal shortage, a larger proportion of high sulfur coals have been mining [1]. Due to combustion temperature, the sulfur(S) content of coal starts to oxidize, and SOx /H_2S gasses are released to the atmosphere through a chimney. Although SOx gasses are made up of sulfur dioxide(SO_2), and sulfur trioxide(SO_3), the SO_3 content is rather low[2]. Fossil fuels are the main cause of the SO_2 emissions in the world, measuring roughly 80% [3]. From the environmental perspective, acidic gas, particularly SO_2 and H_2S, has serious negative effects on the environment (air pollution). This is because combustion of SO_2 and H_2S produces highly toxic byproducts, such as sulfur dioxide, sulfuric acid, sulfurous acid, carbon monoxide [5]. These byproducts affect air quality and also considered as major sources of acidic precipitation, see Kohl and Nielson [4]. Sulfur compounds in the chemical industry and hydrocarbon fueled vehicles also cause corrosion to parts of refineries and combustion engines because of the formation of oxy-acids of sulfur from combustion products [5]. Ma Yuzhen et al [6] considered that the safety critical content of SO_2 was 0.036%, and the amount of SO_2 generated during coal combustion is higher than the safety levels. "The most serious environmental protection law in China history" was implemented since 2015. The new environmental law put forward has higher requirements for FGD. With the development of gas purification technology, organizations have been looking for FGD technology which is more efficient, safe, economical and easy to operate.

2. The desulfurization technology based on environmental safety

The formation of sulfur and its compounds cycle changed through oxidation or reduction reaction in nature. In the course of the development of desulfurization technology, people through chemical or biological methods promote the process of oxidation or reduction reaction, changing gaseous sulfide into liquid, solid sulfides or elemental sulfur and separated, reducing sulfur content in the flue gas. The FGD technology based on the environmental safety is established in sustainable development concept, in the process of reducing the sulfur content of the flue gas, there is no secondary pollution to the environment, no need to deal with the desulfurization slag, liquid, etc., and least carbon emission. Although the removal rate of using conventional technology such as NaOH or CaO to reduce sulfur content in the flue gas is high, the two processes are continuously adding the massive lyes, belonging to the transfer of waste rather than efficient treatment. Biological oxidation desulphurization technology [7] is through the metabolism of sulfur bacteria to convert sulfide into sulfur, such microorganisms include: photosynthetic bacteria, denitrifying bacteria, colorless sulfur bacteria [8].

2.1. Traditional desulfurization technology

The traditional desulfurization technology mainly refers to chemical oxidation method, by adding chemical oxidant to convert sulfide into sulfur. The chemical oxidation widely used in engineering field for a hundred years, the initial process including Burkheiser, Ferrox, Clund and Manchester, Alkali washing method [9], these traditional processes are used less and less because oxidant need to be add constantly, which result in high cost for running. Iron chelate oxidation desulfurization technology is one of chemical oxidation methods widely used. Fe^{3+} has a higher oxidation -reduction potential for turning sulfide to Sulfur. Sulfur can be recycled by coagulation and gravity separation, Fe^{3+} will be regeneration and cyclic utilization [10]. The following shows the chemical process:

$$H_2S_{fluid} + 2Fe^{3+} + L^{n-} \rightarrow 2Fe^{2+} + L^{n-} + 2H^+ + S$$

$$O_{2fluid} + 4Fe^{2+} + L^{n-} + 2H_2O \rightarrow 4Fe^{3+} + L^{n-} + 4OH^- \text{(L as ligand)}$$

In this process, although oxidizer can achieve recycling, the energy consumption is large, the byproduct is more complex, there will be lots of constraints when used in agriculture, such as soil PH, excess soluble salts, aluminum toxicity, sulfite toxicity and etc. [11]. It is also easier to introduce second pollution to the environment.

2.2. Bio-desulfurization technology

The sulfur and its compounds gain or lose electrons by oxidation or reduction reaction, and the morphology of the compound of sulfur can be changed by the cycle. Sulfur of the biological cycle includes the following parts [12], in anaerobic environment conditions and under the action of sulfate reduction bacteria, sulfate is reduced to sulfide. The sulfide is oxidized to elemental sulfur and sulfate under the action of aerobic thiobacillus, which is called the process of biological oxidation. Sulfate is converted to organic sulfur under the action of bacteria, which is called the assimilation and reduction process of sulfur. The biological cycle of these sulfur and the physical and chemical processes are shown in Figure 1. Sulfur circulation is mainly through the biological cycler.

Bio-FGD technology based on biological sulfur cycle theory, it can be divided into three steps. The first step is sulfur dioxide in the flue gas is dissolved by water or absorption tower and converted to sulfite or sulfate, the second step is to use sulfate reducing bacteria to reduce the sulfite and sulfate to sulfide under anaerobic conditions and external carbon sources. The third step is to use the role of sulfur bacteria in aerobic reactor to convert sulfide into sulfur. BioDeSOx process is one of Bio-FGD technologies.

Several companies jointly developed a new bio-desulfurization technology, which was called BioDeSOx process. In this process, comprehensive utilization of alkali absorption and biotechnology

can convert SO_2 to sulfur, alkali also can be regenerated. This not only reduces the raw material adding but also can produce sulfur products to achieve economic benefits [13]. There are no secondary pollutants and emissions produced, therefore it is conducive to the security environment.
SO_2 will be generated in the combustion process of sulfur in coal, which is easy soluble in water and transfer to SO_4^{2-} or SO_3^{2-}[14].

$$SO_2+O^{2-}\rightarrow SO_3^{2-}$$
$$2SO_2+O_2+2O^{2-}\rightarrow 2SO_4^{2-}$$

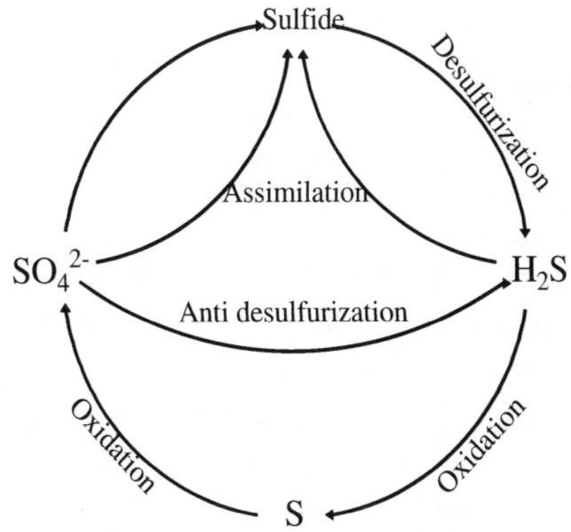

Figure 1. Schematic diagram of sulfur cycle in nature

Liquid phase under anaerobic conditions, the sulfur ion of sulfate and sulfite can gain or lose electrons to form S^{2-}, then the process requires sufficient electron donor to provide electron. Generally, the electron donor can be hydrogen or alcohols (methanol, ethanol), it can also be a high concentration of organic wastewater that the reductive substances COD react with sulfate and sulfite to general S^{2-}.

$$SO_3^{2-} \text{ and } SO_4^{2-}\rightarrow S^{2-}$$

Liquid phase under aerobic conditions, colorless sulfur bacteria transfer S^{2-} to sulfur(S).

$$S^{2-}\rightarrow S$$

3. Biodeso$_x$ bio-fgd technology

In the 1990s, the company Paques began to research and develop the biological desulfurization technology - Thiopaq technology, is mainly used in the removal of H_2S. This has been applied in more than 100 desulfurization projects all over the world, the industrial application has proven that Thiopaq technology is an environmentally friendly and efficient method. Subsequently, this technology created large company's attention in companies such as Shell and UOP [15]. These companies used this technology for removal H_2S in the petroleum and chemical industry. On this basis, the two other big energy companies Hoogovens and Nuon with Paques jointly researched and developed the BioDeSOx technology for removing SO_2 in flue gas. Firstly, SO_2 in the gas will be adsorbed by lye and converted to sulfate, the reduction reaction of sulfate occurs under anaerobic conditions and H_2S would be formed, which is then uses the Thiopaq technology to convert H_2S to S. The application of BioDeSOx technology greatly reduces the running cost and reduces the risk of the secondary environmental pollution. Biological sulfur produced by biological desulfurization has no harmful chemical components, even though it has not pass through purification treatment process can also be widely used in agriculture, which will result in economic benefits [16].

3.1. The process flow

The BioDeSOx technology process as shown in Figure 2 [17].

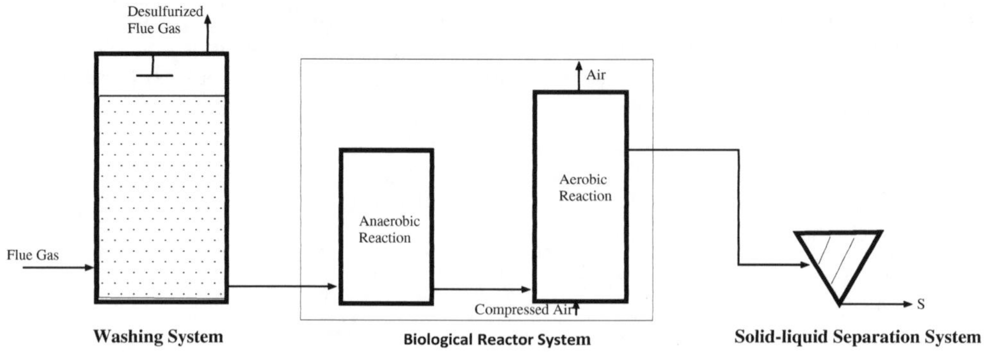

Figure 2. The BioDeSOx technology process

3.1.1. Washing system. In the washing system, gas is pumped into the bottom of the washing tower and discharged from the upper part of the tower. There is a sprinkler system at the top of the tower. When gas goes through washing liquid spray at the top, sulfur oxides, halides and dust are removed from flue gas by the washing liquid and the purified gas is discharged from the top of the tower. The washing liquid that adsorb the sulfur oxides and others by products is discharged from the bottom of the tower to the anaerobic system.

In order to achieve high SO_2 removal efficiency, lye such as NaOH, Na_2CO_3, and Ca $(OH)_2$ was widely used as a washing liquid because of the effective reaction between SO_2 and alkaline substances. In the traditional way of desulfurization, the price of sodium salt (Na^+) is high, companies usual make use of calcium salt (Ca^{2+}), which is cheaper but may easier cause blockage problems. The lye of BioDeSOx biological systems can be renewable and restore the alkaline of washing liquid, the method of sodium salt can be used economically in this system to avoid the blockage problem.

The coal combustion cycle not only contains SO_2, but also contains a large number of CO_2. CO_2 is easily soluble in alkali solution, which can make the circulation system rapidly form a $NaHCO_3$ buffer. The buffer system has a very strong resistance to the fluctuations of the concentration of SO_2 in the flue gas and maintains a stable and low concentration of SO_2 in the final exhaust gas [9].

$$SO_2 + NaHCO_3 \rightarrow NaHSO_3 + CO_2$$
$$SO_3 + 2\ NaHCO_3 \rightarrow Na_2SO_4 + 2\ CO_2 + H_2O$$

The concentration of $NaHCO_3$ in flue gas is balanced with the concentration of CO_2, and maintains the PH at an optimum range required by the biological reaction. There are no need to add NaOH unless the consumption of the washing liquid to remove sulfur is too large.

3.1.2. Biological reactor system. The core patent of the BioDeSOx technology is the biological reactor [18], which is divided into anaerobic system and aerobic system. The anaerobic system is using an IC reactor whereas the aerobic system is using CIRCOX reactor [19].

The detergent containing sulfite and sulfate is pumped into the anaerobic system after a simple precipitation. In the anaerobic system, sulfite and sulfate are converted into sulfides through a microbial process and then pumped into the aerobic reaction system. The aerobic microbial convert the sulfides to elemental sulfur and re-used lye [20].

In the anaerobic system, sulfite and sulfate are converted to sulfide required a donor element to provide electrons. The CODcr concentration of wastewater from coal (generally above 5000mg

/ L), it is a very economical way to use the high concentration organic wastewater as the reducing agent. Considerable testing and analysis must be completed to understand the wasterwater characteristics whether it can use reasonable, then it can greatly reduce the operational cost of the FGD in coal chemical industry [21].

$$NaHSO_3 + 0.75\ HAc \rightarrow NaHS + 1.5\ H_2O + 1.5\ CO_2$$

$$Na_2SO_4 + HAc \rightarrow NaHS + H_2O + CO_2 + NaHCO_3$$

In the aerobic system, sodium sulfide was oxidized to sulfur by colorless sulfur bacteria, as follows:

$$NaHS + 0.5\ O_2 \rightarrow S + NaOH$$

The CIRCOX reactor is a sleeve structure tower where microorganisms in the reactor will form granular sludge and have good separation and settlement properties. Both the wastewater containing sulfide and compressed air enter into reactor from the bottom. The compressed air enters into the inner cylinder to raise the granular sludge in to air where it mixes with the wastewater resulting in aerobic micro-organisms reaction with wastewater. At the top of reactor, compressed air is discharged through the exhaust holes and slurry containing sulfur overflow into the separator. Because there is no disruption airflow, granular sludge with good settlement performance will settle in the bottom of outer cylinder and participate in the next cycle of reaction. Liquid overflow into the solid-liquid separation system, after separating the sulfur, the high pH value of the liquid is returned to the washing tower to realize the lye regeneration. The structure of CIRCOX reactor as shown in Figure 3.

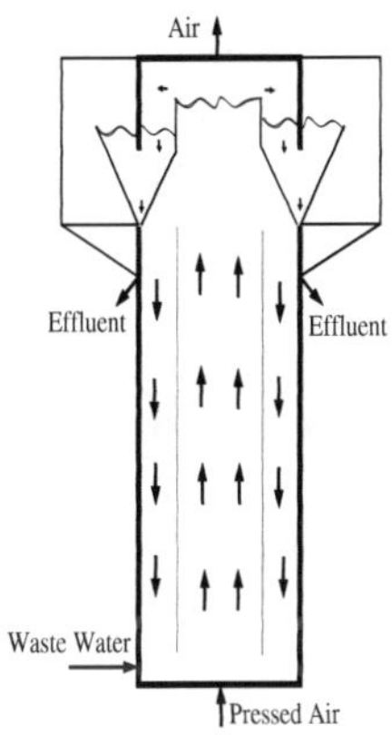

Figure 3. CIRCOX reactor structure diagram

3.1.3. Solid-liquid separation system. Sulfur generated in the aerobic biological reactor and separated by solid-liquid separator. A small portion of sulfur slurry is returned to bioreactor for controlling the concentration of sulfur/ microbial, the vast majority of sulfur removed as the slurry where the solid content is 5-10%. The purity of sulfur is 90%, and the remainder is organic matter and trace amounts of salts. The picture of sulfur separated by solid-liquid separator is shown in Figure 4 [22].

3.1.4. Features. Compared with the traditional desulfurization technology, the characteristics of BioDeSOx technology is obvious:

1) Biological desulfurization technology is simple, reliable, easy to operate and can meet the requirements of desulfurization.

2) Can directly use the local high concentration organic waste water or waste residue as an electron donor, not only meeting the requirements of desulfurization, but also can achieve the aim of treatment and utilization of waste as the circular efficient process.

IOP Conf. Series: Earth and Environmental Science **154** (2018) 012018 doi:10.1088/1755-1315/154/1/012018

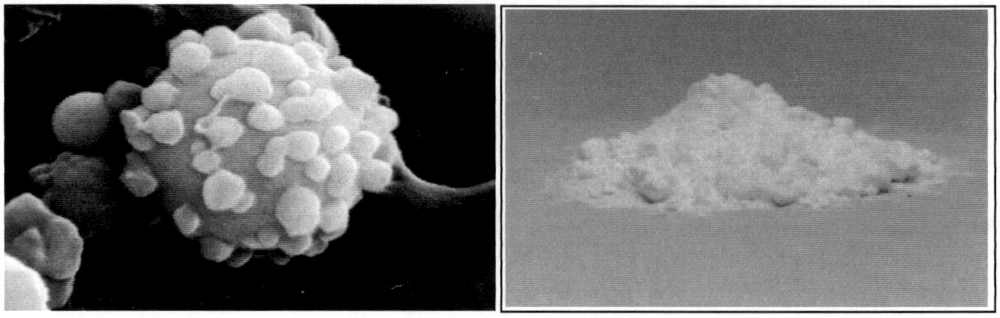

Figure 4. Sulfur separated by solid-liquid separator

3) Have good adaptability to the high CO2/SO2 concentration of gas, will consume less lye, low energy consumption.

4) Biological sulfur has good hydrophily and mobility, to avoid the phenomenon of sulfur blocking and equipment corrosion, biological sulfur can be directly used as raw material for agricultural or compound fertilizer, which increase the economic efficiency.

5) Lye can regenerate and recycling, thus reducing the operational cost of desulfurization process.

6) There is no waste liquid and waste residue discharged, has no the risk of secondary pollution, carbon emissions are low.

The concentration of the SO_2 in the process of coal combustion is low, the fractional volume will generally not exceed 25%. The high ratio of gas CO_2/SO_2 (this kind of gas is not suitable for using traditional sulfur treatment) is an excellent choice for the BioDeSOx desulfurization technology.

3.2. Influence factor

BioDeSOx technology is simple and stable operation, but as a biological system, microorganism survival still requires certain conditions, the main factors are the following.

1) Washing tower requires a certain pH buffer range. Microorganism have a fixed range of sulfur processing load, the sulfur components in intake-air are larger , the washing liquid sulfate fluctuation is then bigger, that will have an impact to the subsequent biological treatment system and thus reducing the processing efficiency. If the CO_2 of the intake-air is within a certain pH value, the formation of HCO_3^- solution with the role of buffer.

2) Oxidation reduction potential in aerobic system. Oxidation reduction potential is decided by the oxygen quantity. If the redox potential is too high, the sulfide is oxidized to sulfate. If it is too low, this will cause the low removal of sulfide, decrease the absorption efficiency of lye and reduce the removal rate of the gas SO_2 and H_2S. The redox potential is an important parameter with an ideal control range of -300mV to -400mV, which can be adjusted by controlling the frequency of the conversion blower.

3) Electrical conductivity. The electrical conductivity reflect the concentration of salt in the reactor. The conductivity is higher, the salinity of liquid in the reactor is higher, which will resulting in the death of microorganisms. The lower the conductivity and lower liquid salinity is not conducive to the growth of microorganisms. The electrical conductivity of the reactor can be controlled by discharging the reactor liquid and adding the softened water.

4) Nutrient additions. Microbial growth requires adding quantitative nutrient salts. The amount of nutrients can be determined according to the amount of sulfur removal. If nutrient additions are too few, microorganisms do not grow and adding too much can cause poisoning. Microbial biomass determines desulfurization effect.

5) Temperature. Microbial growth requires an appropriate temperature. If the temperature is too high, it will cause the death of microorganisms, if the temperature is too low will lead to a decrease in microbial activity. Desulfurizing bacteria suitable temperature range is between 30 degrees to 40 degrees.

3.3. Benefits analysis

By comparing the traditional alkaline cleaning method, iron chelate and BioDeSOx desulfurization technology, one can clearly see the advantages of biological desulfurization technology.

Table 1. Advantages Comparative Analysis of Desulfurization Technolog

	Alkaline cleaning method	Iron chelate method	BioDeSOx method
Chemical cost	RMB20000yuan/tS	RMB5000 yuan/tS	RMB2000 yuan/tS
Operation time	12h/d	12h/d	1h/d
Desulfurization efficiency	>90%	>90%	>98%
External emission	Waste chemicals	Waste chemicals	Non toxic salt
Failure rate	Easy to jam, high failure rate	Easy to jam, high failure rate	Not easy to jam, low failure rate
Economic performance	N/A	N/A	Produce biological sulfur with economic benefits

4. Case analysis

4.1. Project profile

Yixing xielian heat and power plant is a Sino Thai joint venture, located in the Yixing City which is "township of Chinese environmental protection". This city is in the Taihu Lake Basin, which belongs to the key environmental protection area. In 2003 - 2005, the boiler flue gas desulfurization came into operation with the implementation of the 2 x 125MW coal-fired thermal power unit expansion project. BioDeSOx desulfurization technology [23] was chosen for this project. Design of flue gas emissions 1.1×106m^3/h, SO_2 content of the gas is 0.037%. The Yixing xielian company also had a citric acid plant beside the heat and power plant. This is a great amount of reducing agent will be needed in the anaerobic systems of the flue gas desulfurization. The citric acid plant produces a large quantity of high concentration wastewater can be as electron donors providing microelectronic and eventually convert the SO2 in flue gas to sulfur.

The process as shown in Figure 5 mainly includes washing, cooling and precipitation, anaerobic treatment, aerobic treatment system, solid-liquid separation system, etc.

4.2. Benefits Operation Analysis

After the completion of the project, there will be 3 million high methane concentration gas directly into the biogas generating units and the rate of flue gas desulfurization up to 98%, emissions concentration of SO_2 < 100mg / m^3, and removal dust. Because there is no addition of limestone, there is no chemical waste, which avoids the secondary pollution; microorganisms in the reaction exist in nature and do not impact the environment. At the same time, sulfur with high application value is produced. The annual production is estimated at 5000 tons, resulting in good economic benefit to offset costs. When compared with the traditional flue gas desulphurization technology, the initial one-time investment is larger for BioDeSOx technology, the economic benefits is significantly once in operations, Also, in view of the life cycle of this technology, the running cost, the effect of energy saving and emission reduction is undoubtedly better than the traditional desulfurization technology.

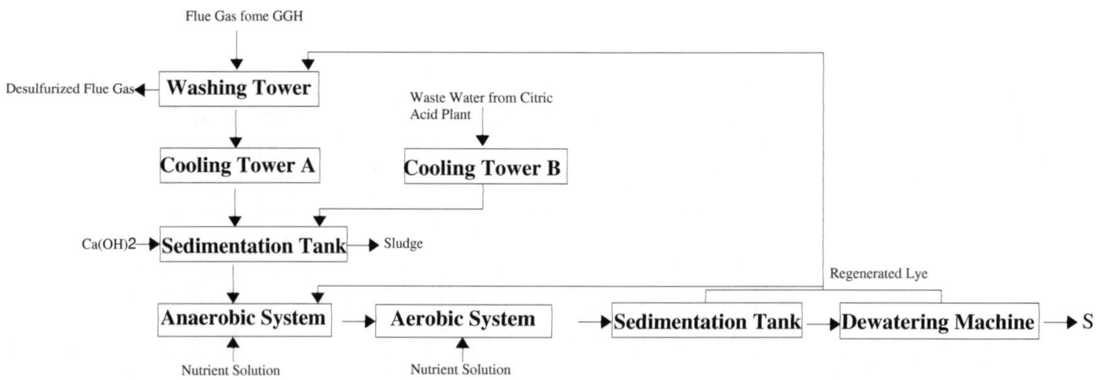

Figure 5. BioDeSOx technology process of Yixing xielian thermal power plant

5. Conclusion

With higher environmental protection standards being striven for, acidic gas generated in the process of coal combustion and coal chemical industry is constantly being improved through energy saving and emission reduction requirements. Bio-desulfurization technology used to convert SO_2 to elemental sulfur with the utilization of recycled high concentration organic wastewater, obviously realizes the lowest environmental pollution and best environmental benefit. The regeneration of alkali liquid will reduce the operational cost. There is no waste gas discharge and energy-saving emission reduction effect is good as well as sulfur resources are produced to realize the additional economic value resulting in good application for the foreseeable future.

6. References

[1] Y. Hu, B. S. Zhen, M. S. Wang, R. B. Finkleman. 2005. Distribution characteristics and origin of sulfur in Chinese Coals. Coal conversion. 28（4）, 1-6.
[2] M. Balat, 2008. Coal-fired power generation: proven technologies and pollution control systems. Energy Sources, PartA30, 132–140.
[3] Osman Kilic; Bora Acarkan; Selim Ay. 2013. FGD investments as part of energy policy: A case study for Turkey. Energy Policy. 62, 1461-1469.
[4] A. Kohl, R. Nielsen 1997. Gas purification. Houston TX: Gulf Publishing Company.
[5] A. K. Gupta, S. Ibrahim. A. Al Shoaibi. 2016. Advances in sulfur chemistry for treatment of acid gases. Progress in Energy and Combustion Science, 54:65–92.
[6] Y. Z. Ma, M. X. Tao, X. M. Zhang, 2008. Hazard of H2S and SO2 Contents in Coalbed Gas and Natural Gas. Geological journal, 82（10）, 1-7.
[7] Y. B. Zheng, W. Wang, 2006. The possibility of application of bio-desulfurization technology in the field of coal chemical industry. Coal Chemical Industry, 123（2）,54-56.
[8] M. C. Hu, T. R. Long, Z. J. Li, 2005. New progress in the research of biogas desulfurization technology. China Biogas, 23（1）,17-20.
[9] Jackso, Yu. 1998. An Advanced Method of Hydrogen Sulfide Removal from Biogas. Sci Industr Res, (16c), 98—103.
[10] K. Christian. T. Frderic, 1998. Waste Gas Biotreatment Technology. Chemical Technology & Biotechnology, 72:303— 319.
[11] R. B. Clark, K. D. Ritchey, V. C. Baligar. 2001. Benefits and constraints for use of FGD

products on agricultural land, Fuel. 80:821-828

[12] P. N. L. Lens, J. G. Kuenen. 2001. The biological sulfur cycle: novel opportunities for environmental biotechnology. Water Science & Technology. 44(8):57-66

[13] G. Cai, 2010. Overview of biological desulfurization technologies for low SO_2 concentrations off-gases. Sulphuric Acid Industry, 6: 44— 47.

[14] Y. Xiang, Z. Wang, Xu,Ch. 2011. Impact of SO_2 concentration on the corrosion rate of X70 steel and iron in water-saturated supercritical CO_2 mixed with SO_2. The Journal of supercritical Fluids, 58(2), 286— 294.

[15] C. Cline, A. Hoksberg, R. Abry, 2003. Biological Process for H_2S Removal from Gas Streams: The Shell-Paques/THIOPAQ™ Gas Desulfurization Process. Paper for the LRGCC, 23-26.

[16] V. Zessen, 2004． E; // British Sulphus Events Sulphur Conference．

[17] X. D. Jin, G. B. Liu, 2010. Characteristic of Flue Gas Biodesulfurization Technology. Environmental science and technology, 23(2), 31-34

[18] Ico, van. den. Born. Donald, C. heng. K. evin. 2005. Bio-desulfurization technology for gas-liquid material handling. Gas Purification, 5 (4), 80 – 82.

[19] Y. L. Zhang, W. B. Hu, Z. Z. Tang, 2002. Application of THIOPAQ technology in acid gas processing plant of refinery. Hangzhou Chemical Industry, 32 (4), 25 – 27.

[20] J. M. Wang, 2009. Shell- Paques bio-desulfurization technology and its application. Chemical Tecnology Market, 32 (11). 29 - 32.

[21] E. Derek. Chitwood. Joseph. S. Devinny. 2001.Treatment of mixed hydrogen sulfide and organic vapors in a rock medium biofilter. Water Environment Research. 73(4):426-434.

[22] K. J. Wang, C. Hu, 2016. Biological sulfur cycle and new development in desulfurization technology. Environmental protection, 1B, 69-89

[23] M. B. Chen, M. F. Liao, L. N. Huang, C. J. Qin, 2011.Application of bio-desulfurization technology in thermal power plant of Yixing, Energy environmental protection, 6 (6), 37 - 40.

Acknowledgements

This research was financially supported by the State Key Research Development Program of China (2016YFC0801402、2016YFC0600708).

IOP Conf. Series: Earth and Environmental Science **154** (2018) 012019 doi:10.1088/1755-1315/154/1/012019

Environmental impact assessment of european non-ferro mining industries through life-cycle assessment

Shahjadi Hisan FARJANA [1], Nazmul HUDA [2], M. A. Parvez MAHMUD [3]

[1] Ph.D. candidate, School of Engineering, Macquarie University, Sydney, Australia
[2] Lecturers, School of Engineering, Macquarie University, Sydney, Australia
[3] Ph.D. candidates, School of Engineering, Macquarie University, Sydney, Australia
E-mail: nazmul.huda@mq.edu.au

Abstract. European mining industries are the vast industrial sector which contributes largely on their economy which constitutes of ferro and non-ferro metals and minerals industries. The non-ferro metals extraction and processing industries require focus of attention due to sustainability concerns as their manufacturing processes are highly energy intensive and impacts globally on environment. This paper analyses major environmental effects caused by European metal industries based on the life-cycle impact analysis technologies. This research work is the first work in considering the comparative environmental impact analysis of European non-ferro metal industries which will reveal their technological similarities and dissimilarities to assess their environmental loads. The life-cycle inventory datasets are collected from the EcoInvent database while the analysis is done using the CML baseline and ReCipe endpoint method using SimaPro software version 8.4. The CML and ReCipe method are chosen because they are specialized impact assessment methods for European continent. The impact categories outlined for discussion here are human health, global warming and ecotoxicity. The analysis results reveal that the gold industry is vulnerable for the environment due to waste emission and similar result retained by silver mines a little bit. But copper, lead, manganese and zinc mining processes and industries are environment friendly in terms of metal extraction technologies and waste emissions.

1. Introduction

The European Union metals and minerals industry is an important producer of chromium, copper, lead, silver, and zinc. Europe is rich in metal mining resources where the investment and the production is increasing. In last 5 years, 13 mew mines are opened in the sector of nickel, gold, cobalt, copper, tungsten and zinc mining sector. The challenges in the European mining industries are cost of energy, skills, infrastructure development, technologies and export revenue. Despite the economic benefits, mining industries are responsible for a major share of environmental impacts and global warming effects experiencing by the European subcontinent. Wastes generated from the mining extraction processing steps are one of the leading causes of significant environmental effects. The non-ferro metal mining industries are most of the countries from European union. For example, copper is found and produced in Bulgaria, France, Poland and Russia. Gold is obtained from Russia, Sweden, Bulgaria and Finland. Lead is produced from Ireland, Macedonia- the former Yugoslav Republic of, Poland and Sweden. Manganese mines are distributed in Ukraine, Bulgaria, Greece and Bosnia and Herzegovina. Silver is produced in Poland, Russia Sweden and Greece. And lastly zinc mines are located and running in Ireland, Finland, Poland, Portugal, Russia and Sweden. So, it can be easily inherent that non-ferro mining industries are prevalent everywhere in Europe. Even though these mining industries are in different countries within the European continent, they share same technologies or electricity generation loads which made it easy to discuss them in the same platform. Due to the position and reserve of non-ferro metal mines in European countries, they are the focus of attention in terms of environmental sustainability [1]-[5].

2. Non-ferro mining industries and sustainability concerns

Among numerous non-ferro mining industries, few well known and easily found non-ferro mining industries are chosen for the analysis here in this paper. These are copper, gold, lead, manganese, silver and zinc. These metals and their sustainability values are discussed here shortly in brief.

2.1. Copper

Copper is a soft, malleable and ductile metal which is a high thermal and electrical conductive metal. Copper is used as building material and for making alloys. It is a naturally occurring stable metal to be used in the same form. The main application of copper is in the field of electrical wire, industrial machinery, and plumbing and electronics devices. In terms of sustainability, when copper gets mixed with aquatic environment, it is harmful for all marine species and plants, which is toxic and can hinder the survival and growth of the plants in extreme levels. High levels of copper exposure are detrimental for human health in terms of toxicity which can cause nausea, vomiting and diarrhoea, liver damage even death. Copper mining is also responsible for deforestation, land degradation and water pollution [3], [4], [8], [10], [12].

2.2. Gold

Gold is a bright, soft, malleable and ductile metal which found in a solid solution in combination with silver. Due to its alloying capability, gold is widely used to make jewellery by polishing and colouring. Among other applications, gold metallic compounds are used for making medicine, electronics. In case of environmental impact, gold mining effects largely due to its large amount of wastes released from the mining sites. From open-pit mining and heap leaching process, gold mines release a huge amount of toxic waste per year. These toxic wastes easily get mixed with nearby soil or water to destroy the environment. Moreover, due to underground rock distributed over the mining, gold mines are responsible for acid mine drainage. Use of mercury in the gold mining process thus increases the pollution [1], [6], [7], [9], [13], [14].

2.3. Lead

Lead is a soft malleable denser metal which is post-transition metal. Due to the mechanical properties like low melting point, lead metals are used to make bullets. It is also used to make the architectural metals. The extraction and production technologies of lead mining have significant impact on soil and environment. Accumulation of lead in the plants and soils can remain in the same form for hundreds of years without any decay which can hinder their reproduction and development [2], [11].

2.4. Manganese

Manganese is a free element which is found in combination with iron in nature and is a free element. It is an essential element for iron and steel production due to its alloying and subsidizing properties. Manganese is also used in alloying aluminium. Mining of manganese is highly impactful for soil, water and human health [5], [17].

2.5. Silver

Silver is a lustrous, highly conductive, transition chemical element. Silver has high electrical conductivity and mechanical conductivity. Silver is a highly reflective precious metal which is mostly used for making jewellery and silverware. Silver is also used for wound dressing and making electronic devices. In terms of sustainable mining, silver mining causes erosion which contaminates soil water, land and groundwater as well. It also leads to the formation of sinkhole and loss of biodiversity [1], [6], [7].

2.6. Zinc

IOP Conf. Series: Earth and Environmental Science **154** (2018) 012019 doi:10.1088/1755-1315/154/1/012019

Zinc is one of the most abundant chemical element found is earth which is like magnesium due to chemical properties. Zinc is widely used as abrasive agent and used in batteries. Zinc is an alloying element. Zinc compounds with variable solubility gets mixed with the environment during extraction and processing in zinc mining, which make further reactions while meets soil, water and environment. Zinc mining is also harmful for human health due to metallic compound emission [11], [15], [16].

Table 1. Life-cycle impact analysis results-human health.

	Label	Copper	Gold	Lead	Manganese	Silver	Zinc
CML	Human toxicity	0.0321	100	0.0006	0.0005	0.4546	0.0026
ReCipe	Global warming, Human health	0.0048	33.9909	0.0029	0.0097	0.2482	0.0122
	Ozone formation, Human health	0.0176	92.3299	0.0043	0.0154	0.7228	0.0178
	Human carcinogenic toxicity	0.0228	100	0.001	0.0008	0.3908	0.0025
	Human non-carcinogenic toxicity	0.0932	73.7804	0.0052	0.0027	0.7627	0.599
	Water consumption, Human health	0.1277	100	0.002	0.2435	2.0986	0.0082

Table 2. Life-cycle impact analysis results-ecotoxicity.

	Label	Copper	Gold	Lead	Manganes e	Silver	Zinc
CML	Fresh water aquatic ecotox.	0.0227	100	0.0005	0.0006	0.3718	0.0025
ReCip e	Freshwater ecotoxicity	0.0283	100	0.0009	0.0017	0.4201	0.0175
	Marine aquatic ecotoxicity	0.0225	100	0.0005	0.0006	0.3719	0.0026
	Marine ecotoxicity	0.1049	80.4767	0.0045	0.0024	0.8611	0.5461
	Terrestrial ecotoxicity	0.0182	100	0.0024	0.007	0.2267	0.0329
	Terrestrial ecotoxicity	0.2382	18.7327	0.003	0.0065	1.7905	0.0889

3. Life-cycle assessment

Life cycle assessment is a powerful tool to assess the environmental effects caused by a manufacturing process throughout its whole life. The type of life cycle assessment could be from cradle to grave which includes each input materials, output emissions and waste emissions. Life cycle assessment can be of cradle to gate which only considers the input materials and output final product up to the gate of manufacturing factory. The life cycle assessment consists of 4 individual steps: goal and scope definition, life-cycle inventory or data collection, life-cycle impact assessment or environmental effect analysis and interpretation and recommendation of the inventory and results. At the first stage, the goals of the particular life cycle assessment are clarified for better understanding of the producers and consumers according to ISO 14040. This stage defines the products and processes with informative description of the assessment. The system boundary and software, methodologies and allocation techniques are also discussed. In the second step, life cycle inventory analysis is made to identifies, quantifies and report all the input materials, output emissions, output products, waste emissions according to ISO 14041. Allocation on the basis of mass, economics or causal basis also required to be mentioned. In the third step, life cycle impact assessment is carried out according to ISO 14042. The methodologies followed for the assessment procedure should be based on ISO standard. There are many standard methodologies based on different regional and geographical context. For example, in the Oceania there is Australian Indicator method. For European geographic context, there are CML and ReCipe method. For North American region there is TRACI method. According to the methodology, the impacts on land, water, air, ecosystem or human health are assessed. The analysis results can be presented in their exact form or can be presented using characterization or normalization. There are optional or additional results like normalization, grouping or weighting of the indicator results and data quality analysis techniques. These elements are–Category indicator and classification, Characterization, Normalization and Valuation/weighing using ISO 14043. The goal of this research work is to analyses the cradle-to-grave life-cycle environmental effects of non-ferro metals like copper, gold, lead, manganese, silver and zinc where the mining inventory datasets are collected from EcoInvent database focusing on European continent to access their impacts on human health, ecosystems and global warming under various categories. The lifecycle impact assessment is done using SimaPro software version 8.4.. The analysis methods are chosen as the Recipe method and the CML method. These methods are the most common ones used for life-cycle impact assessments related to mining and mineral-extraction activities [5]-[14].

4. Results and discussion

Table 1, 2 and 3 illustrates the analysis results obtained from the SimaPro software. In terms of human health, ecotoxicity and global warming, gold mining and extraction processing routes contribute largely compared with other non-ferro metal commodities considered in this research. The reason should obviously be the large amount of wastes released during the gold extraction in open-pit mining and heap leaching process. All the other mines show similar and non-major impact assessment results rather than gold mining process in European continent. Similar results are found from both CML and ReCipe method. This similarity also clarifies the consistency among the life-cycle impact assessment methods towards the mineral industries routes.

Table 3. Life-cycle impact analysis results-global warming.

Label	Copper	Gold	Lead	Manganese	Silver	Zinc
Global warming (GWP100a)	0.0048	33.8841	0.0029	0.0097	0.2466	0.0121
Global warming, Human health	0.0048	33.9909	0.0029	0.0097	0.2482	0.0122
Global warming, Terrestrial ecosystems	0.0048	33.9906	0.0029	0.0097	0.2482	0.0122
Global warming, Freshwater ecosystems	0.0048	33.9907	0.0029	0.0097	0.2482	0.0122

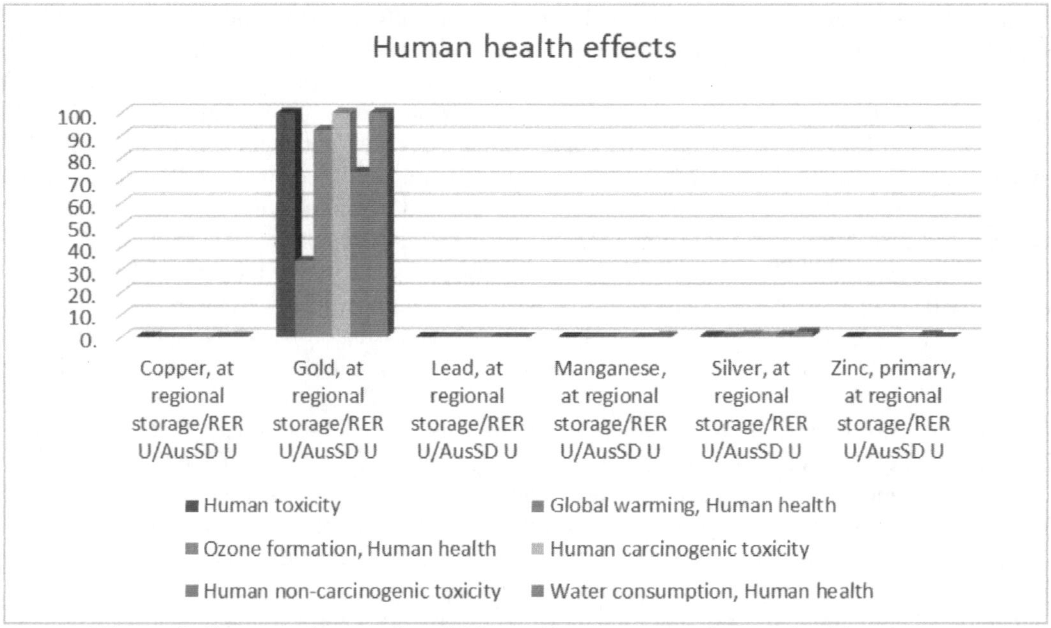

Figure 1. Comparative results of life-cycle assessment- human health category.

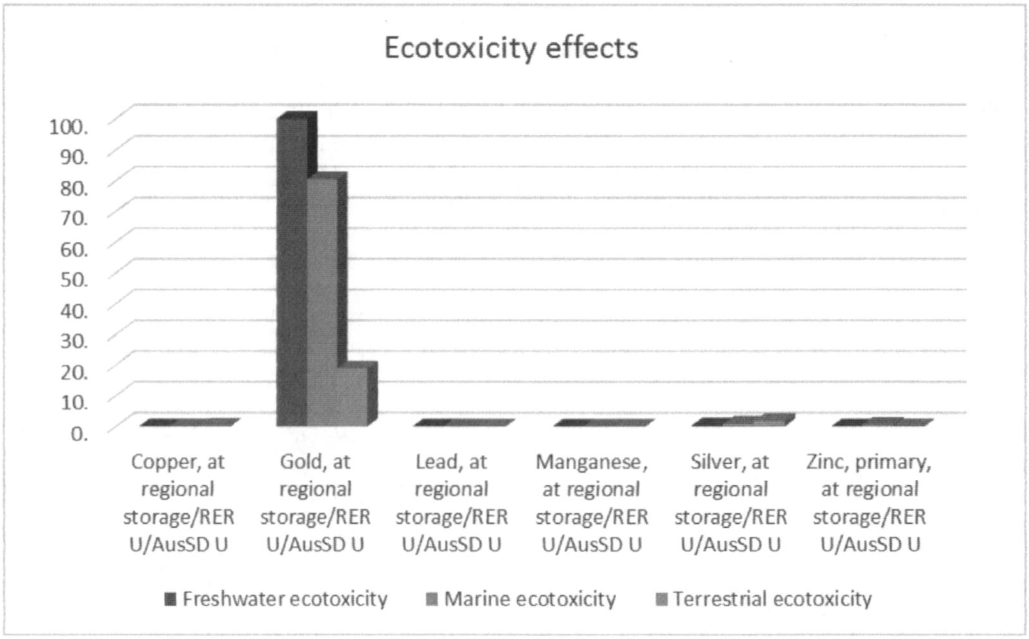

Figure 2. Comparative results of life-cycle assessment- ecotoxicity category.

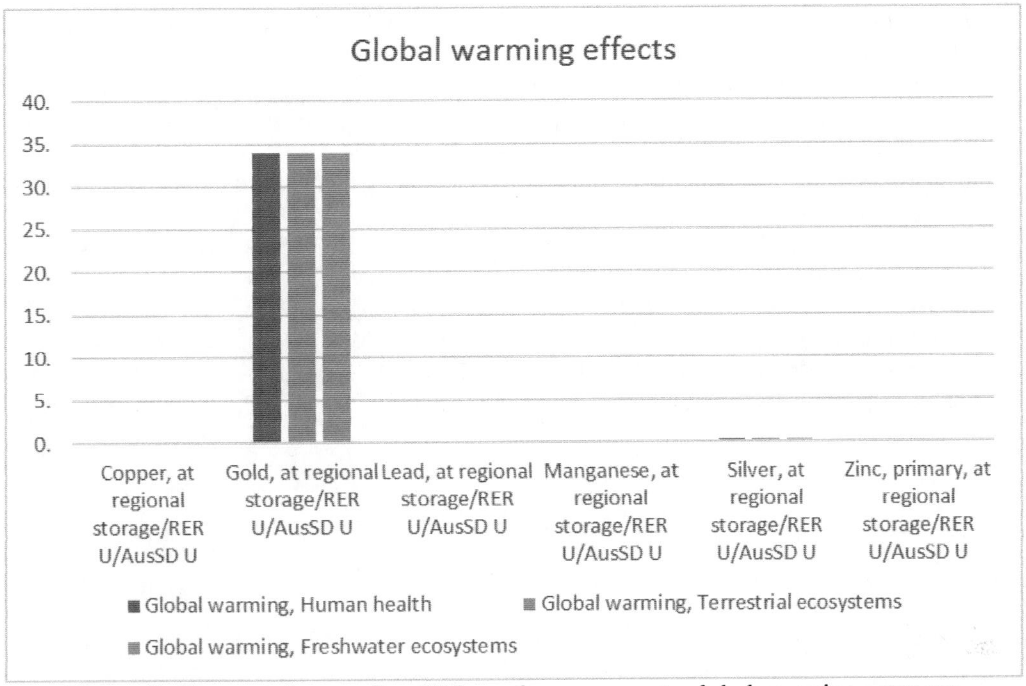

Figure 3. Comparative results of life-cycle assessment- global warming category.

5. Conclusion

In conclusion, this research work assesses the environmental effects caused by the leading non-ferro mineral industries in European continent. The major categories under consideration were global warming, ecotoxicity and human health using CML and ReCipe method. The results reveal that the gold industry is effecting the whole environment largely rather than other non-ferro mining industries. This paper indicates that the largest contributor to global warming and greenhouse gas emissions is gold mining due to their large amount of waste release towards the environment. Countries where economy is strongly dependent on mining industries, especially on gold-mining, should focus on to their key manufacturing technologies to protect their own environment and country from the negative effects of global warming both on their resources and human health. Effective measures should be taken to control the waste emissions from gold industries to reduce the effects.

6. References

[1] M. Ackley. 2008 Evaluating Environmental Risks in Mining: A perceptual study at the Vatukoula gold mine in Fiji 180
[2] L. C. Assessment, International P E and Inventory L C Lead Metal Production LCA
[3] J. Castro-Molinare. 2013 Sustainability analysis of copper extraction and processing using LCA methods **53** 243
[4] K. U. Copper. Environmental Profile
[5] IMnI 2014 Lifecycle Assessment of Global Manganese Alloy Production
[6] W. W. Ingwersen. 2010 Advances in life cycle assessment and emergy evaluation with case studies in gold mining and pineapple production 1–134
[7] W. W. Ingwersen. 2011 Emergy as a Life Cycle Impact Assessment Indicator: A Gold Mining Case Study *J. Ind. Ecol.* **15** 550–67
[8] J. Kelly, Q. Dai, A. Elgowainy, S. A. Group and E. S. Division 2015 Updated Life Cycle Inventory of Copper : Imports from Chile
[9] D. Korpela. 2014 Costa Rica Crucitas gold mining project ESIA
[10] D. Kupferinstitut, L. Cycle, C. Am, G. Tel and A. Klassert 2002 Life Cycle Assessment of Copper Products *Int. J. Life Cycle Assess.* **7** 156–63

[11] S. Nazari and M. Gharabaghi. 2015 Investigation on life cycle assessment of lead and zinc production *Int.Journal Min. Geo-Engineering* **49** 245–52
[12] T. E. Norgate. 2001 A Comparative Life Cycle Assessment of Copper Production Processes
[13] Norgate T and Haque N 2012 Using life cycle assessment to evaluate some environmental impacts of gold production *J. Clean. Prod.* **29–30** 53–63
[14] P. K. Tsang. 2014 Sustainable Gold Mining- Life Cycle Assessment of Cyanidation and Thiosulphate Leaching
[15] E. Van Genderen, M. Wildnauer, N. Santero and N. Sidi 2016 A global life cycle assessment for primary zinc production *Int. J. Life Cycle Assess.* **21** 1580–93
[16] M. Werder and A. Steinfeld 2000 Life Cycle Assessment of the Conventional and Solar Thermal Production of Zinc and Synthesis Gas *Energy* **25** 395–409
[17] L. A. Westfall, J. Davourie, M. Ali and D. McGough 2016 Cradle-to-gate life cycle assessment of global manganese alloy production *Int. J. Life Cycle Assess.* **21** 1573–9.

IOP Conf. Series: Earth and Environmental Science **154** (2018) 012020 doi:10.1088/1755-1315/154/1/012020

Initial solubility & density evaluation of Non-Aqueous system of amino acid salts for CO_2 capture: potassium prolinate blended with ethanol and ethylene glycol

Ghulam Murshid[1,*], Sahil Garg[1]

[1,*]Department of Petroleum & Chemical Engineering, Sultan Qaboos University, Muscat, Oman

Abstract. Amine scrubbing is the state of the art technology for CO_2 capture, and solvent selection can significantly reduce the capital and energy cost of the process. Higher energy requirement for aqueous amine based CO_2 removal process is still a most important downside preventive its industrial deployment. Therefore, in this study, novel non-aqueous based amino acid salt system consisting of potassium prolinate, ethanol and ethylene glycol has been studied. This work presents initial CO_2 solubility study and important physical properties i.e. density of the studied solvent system. Previous work showed that non-aqueous system of potassium prolinate and ethanol has good absorption rates and requires lower energy for solvent regeneration. However, during regeneration, solvent loss issues were found due to lower boiling point of the ethanol. Therefore, ethylene glycol was added into current studied system for enhancing the overall boiling point of the system. The good initial CO_2 solubility and low density of studied solvent system offers several advantages as compared to conventional amine solutions.

1. Introduction

In chemical absorption processes, amines such as monoethanolamine (MEA), diethanolamine (DEA) and methyldiethanolamine (MDEA) are the most common choice as chemical solvents that have been used industrially for acid gas removal [1]. The main advantage of this technology based on chemical absorption of CO_2 by aqueous amines is its great efficiency; more than 90% of low pressure CO_2 can be captured because of the fast acid-base reaction. However, this process has not yet demonstrated to be suitable for the application to CO_2 removing from large scale commercial power plants (up to 500 ton CO_2/h), due to energy required for absorbent regeneration [1]. Besides the regeneration energy, other disadvantages are large equipment size [2], solvent volatility [3], solvent toxicity [4], solvent degradation [5]-[7], and high equipment corrosion [8], [9]. In order to make more attractive implementation of the CCS technology to large scale fossil-fuelled power plants, considerable research efforts are undertaken from engineering and chemical point of view, to make the process more economical efficient as well as environmental safe. To achieve this objectives, the absorbent efficiency and its thermal and oxidative stability should be increased, as well as energy costs of the absorbent regeneration should be decreased while maintaining (or enhancing) the other attributes.

To avoid the issues of solvent volatility, toxicity, degradation and corrosion, several researchers have come up with a new class of solvents named as amino acid salts (AASs) [10]-[21]. AASs are the salt form of natural amino acids. They are therefore formed by neutralizing an amino acid compound with an alkali metal hydroxide. Thus, the alkalinity of the amino group is increased, i.e. the amino acid reacts selectively with acid gases such as CO_2. Due to their ionic nature, both AASs and their absorption products are conveniently non-volatile [11], [22]. As reported by Eide-haugmo et al. [23], AASs have a high biodegradability and their toxicity is an order of magnitude lower than any of the other alkanolamines, piperazine (PZ) or ammonia. Moreover, Fischer observed that AASs such as potassium N, N-dimethylglycinate, taurate and sarcosinate demonstrated lower degradation rates and less volatile degradation product emissions in comparison to alkanolamines such as MEA and DEA [24]. However, it was recently pointed out that the regeneration energy demand is one of main research area where AASs might have to advance [25]. Evidently, there is a great need for more fundamental research into new chemical solvents.

Earlier, alkanolamines were also used in non-aqueous solutions [26]. An example of this is the Sulfinol process using a mixture of diisopropylamine (DIPA), sulfolane (tetrahydrothiophene dioxide), and water. This process had shown the capability of removing carbonyl sulfide (COS) and mercaptans together with H_2S and CO_2 [26]. And recently, several researchers have investigated that non-aqueous amine/alcohol solvents offer various improvements such as low binding energy, low heat capacity and high cyclic loading, resulting in a significant decrement of regeneration energy demand of CO_2 absorbed solvents [27]-[29]. Lately, Shen et al. [30] investigated absorption rate of CO_2 in non-aqueous system of potassium prolinate/ethanol (ProK-EthOH) solution. It was also examined that the ProK-EthOH solvent showed faster absorption rate than aqueous ProK solvent and as well as lower regeneration temperature. However, solvent loss especially ethanol loss during regeneration of solvent was the major issue due to low boiling point of the ethanol need to be addressed for the future studies [30]. Therefore, in this work, ethylene glycol (EG) was mixed with ProK-EthOH solution, so as to increase the boiling point of the azeotropic mixture of EthOH and EG. Moreover, the data on absorption performance of non-aqueous AASs system is scarce in available literature. Here, this work is focused on evaluating the initial CO_2 solubility study of the developed non-aqueous system and evaluating the density of the system.

Knowledge of the physical property such as density of non-aqueous system of ProK-EthOH-EG is essential for process design of gas treating units [31]-[33]. Solution density is important for the mass transfer rate modeling of absorption and regeneration because these properties influence the values of the liquid side mass transfer coefficient [34], [35]. Therefore, in this study, physical properties were measured over the temperature range of 298.15 to 343.15 K and at ProK mass fractions of (0.05, 0.10, 0.15, 0.20, 0.25 and 0.30) with the remaining solution containing equal mass fractions of EtOH and EG. In addition, initial CO_2 solubility is investigated for all concentrations into the studied solvent at 5 bar initial pressure.

2. Experimental section

2.1. Chemicals and solution preparation

The chemicals such as l-proline (Pro), ethanol (EthOH) and ethylene glycol (EG) were provided by Merck Chemicals Germany and potassium hydroxide (KOH) was supplied by Aladdin Reagent China and were used as such. However, the gases such as carbon dioxide (CO_2) and nitrogen (N_2) were provided by Muscat gases. In order to prepare potassium prolinate/ethanol (ProK-EthOH) solutions, equimolar mixture of Pro and KOH was added into equal masses of EthOH and EG. All the chemicals were mixed at 400 rpm and atmospheric pressure using a magnetic stirrer for 2 to 3 hrs till a clear and uniform solution was formed. The weight measurements of the chemicals (Pro, KOH, EthOH and EG) were performed using an analytical balance (Mettler Toledo) with a precision of ± 0.1 mg. All the prepared solutions were preserved in a tightly closed glass container so as to minimize the exposure of air for any contamination. Six different concentrations (0.05, 0.10, 0.15, 0.20, 0.25 and 0.30) of ProK were prepared in EthOH-EG solution in terms of mass fraction. All physical properties were measured in a temperature range of 298.15 to 343.15 K, at every 5 K interval.

2.2. CO_2 absorption experiments

In this study, initial CO_2 solubility tests were performed in order to check the absorption capacity of the newly formulated system of ProK-EthOH-EG. For this purpose, a high pressure solubility cell was used. A brief description of this solubility cell was presented in our previous papers [20], [21]. CO_2

solubility in ProK-EthOH-EG system were measured using a pressure drop technique [36] at 313.15 K. The method of CO_2 loading calculation was described briefly in literature [21], [30], [36].

2.3. Density measurements
The density of ProK-EthOH-EG solutions was measured at atmospheric pressure using a density meter (Anton Paar DMA 4500 M), with a repeatability of ± 5×10^{-5} g.cm^{-3} and temperature accuracy of ± 0.01 K. The equipment works on the theory of oscillation U-tube method [37], [38]. Before any measurements, the cell of the density meter was extensively cleaned with ethanol and water of Millipore quality and dried with a built in air blower. The cleaning and drying procedure was done for the reliability and accuracy of the measured data. The equipment was calibrated for the whole measurement range with water of Millipore quality. In a similar way, the density of the ProK-EthOH-EG solutions was measured. Each density measurement was replicated three times and the calculated uncertainty in the density and temperature data was ± 6×10^{-5} g.cm^{-3} and ± 0.02 K respectively.

3. Results and discussion

3.1. Initial CO_2 solubility test
In this study, initial solubility of ProK-EthOH-EG system was examined at 5 bar initial pressure and at 313.15 K. The obtained solubility data was calculated in terms of CO_2 loading which is mol of CO_2 per mol of solvent used and are presented in Table 1. It was observed from table that as the concentration of ProK in solution increases CO_2 loading decreases. Because with the increase in concentration of ProK, the rate of number of CO_2 absorbed in the solution is less than the rate of number of moles of overall solvent and thus causing the loading to decrease. A similar behaviour in CO_2 loading data with increase in concentration of solvent can be observed in literature also [20], [21], [39], [40].

Table 1. Experimental initial CO_2 solubility data in ProK-EthOH-EG solutions at 313.15K.

Concentration of ProK in solvent (w)	Pressure (bar)	CO_2 loading (mol CO_2/mol of solvent)
0.05	3.92	1.456
0.10	3.71	1.310
0.15	3.53	1.196
0.20	3.40	1.079
0.25	3.32	0.965
0.30	3.24	0.871

Table 2. Experimental density data of ProK-EthOH-EG solutions at 0.1 MPa.

T/K	density (*ρ*/g.cm^{-3})					
	ProK (w)					
	0.05	0.10	0.15	0.20	0.25	0.30
298.15	1.3985	1.4025	1.4066	1.4115	1.4187	1.4225
303.15	1.3972	1.4011	1.4051	1.4103	1.4161	1.4197
308.15	1.3962	1.4003	1.4034	1.4095	1.4143	1.4185
313.15	1.3968	1.4008	1.4038	1.4102	1.4149	1.4190
318.15	1.3983	1.4029	1.4057	1.4123	1.4170	1.4203
323.15	1.4021	1.4062	1.4095	1.4146	1.4208	1.4244
328.15	1.4070	1.4102	1.4135	1.4185	1.4260	1.4299
333.15	1.4109	1.4148	1.4173	1.4229	1.4299	1.4338
338.15	1.4123	1.4176	1.4231	1.4268	1.4314	1.4355
343.15	1.4131	1.4186	1.4242	1.4285	1.4321	1.4363

3.2. Density

The experimental density data of ProK-EthOH-EG solutions was measured at several mass fractions (0.05, 0.10, 0.15, 0.20, 0.25 and 0.30) and temperature ranging from 298.15 to 343.15 K. The measured density data are reported in Table 2 and the effect of concentration and temperature on density is shown in Fig. 1. It can be observed from Fig. 1 that density increases as the concentration of solvent increases isothermally, hence showing the effect of concentration. Because, as the concentration of ProK molecules increase in the solutions, the intermolecular interactions between ProK and EthOH-EG molecules might increase due to hydrogen bonding and Van der Waals dispersion forces between them, which in turns increases the density. However, density data decrease linearly with the rise in temperature of the solution at any constant concentration. Because at higher temperatures, the spaces between the molecules get increased resulting in the expansion of the solution volume and thus decreasing the density. Similar behaviour in the variation of the density data has been observed in literature also [37], [41]-[43].

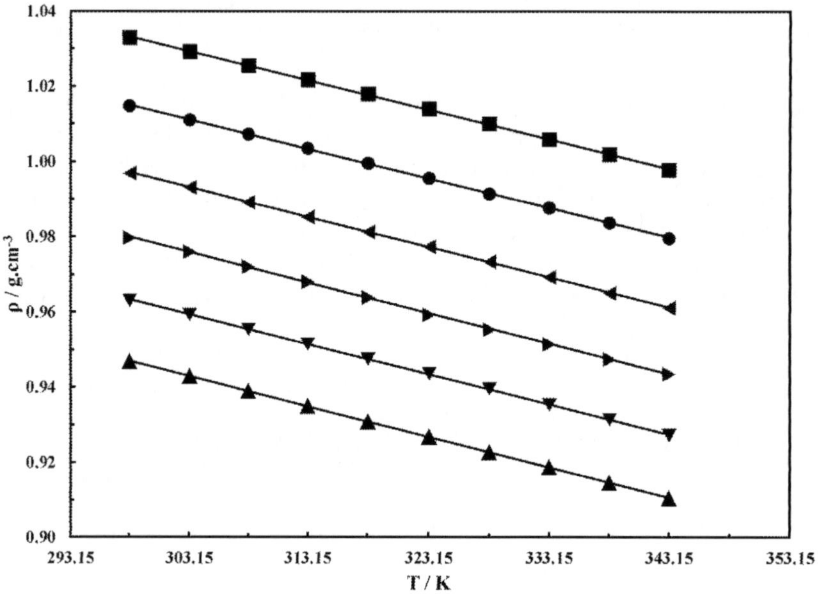

Figure 1. Plot of experimental density versus temperature for several mass fractions of ProK-EthOH-EG solutions; ▲, 0.05; ▼, 0.10; ▶, 0.15; ◀, 0.20; ●, 0.25, ■, 0.30.

4. Conclusion

Experimental density and initial CO_2 solubility data of ProK-EthOH-EG solutions were investigated at several concentrations (0.05, 0.10, 0.15, 0.20, 0.25 and 0.30) and temperatures (298.15 to 343.15 K). The density data increase with increase in concentration of ProK in the solution, while, an opposite behavior was observed with the rise in temperature of the solution. CO_2 solubility decreases with increase in concentration of ProK in the solution. The data obtained in this work could be useful for the designing of efficient CO_2 absorption system. Additional research on the studied solvent is required in terms of reaction mechanism and absorption rate study for the proper understanding before it can be used on a commercial scale for CO_2 removal processes.

5. Acknowledgments

The authors gratefully acknowledge the financial and technical support for this research work by Sultan Qaboos University.

6. References

[1] A. L. Kohl, Nielsen R. 1997 *Gas purification*: Gulf Professional Publishing.
[2] G. T. Rochelle. 2009 Amine Scrubbing for CO_2 Capture *Science* **325(5948)** 1652-1654.
[3] T. Nguyen, M. Hilliard and G. T. Rochelle. 2010 Amine volatility in CO_2 capture *Int. J. Greenhouse Gas Control* **4(5)** 707-715.

[4] B. Thitakamol, A. Veawab and A. Aroonwilas 2007 Environmental impacts of absorption-based CO_2 capture unit for post-combustion treatment of flue gas from coal-fired power plant *Int. J. Greenhouse Gas Control* **1(3)** 318-342.

[5] J. D. Davis 2009 Thermal degradation of aqueous amines used for carbon dioxide capture. *PhD thesis* The University of Texas at Austin, Austin, TX, United States.

[6] C. Gouedard, D. Picq, F. Launay and P. L. Carrette. 2012 Amine degradation in CO_2 capture. I. A review *Int. J. Greenhouse Gas Control* **10.** 244-270.

[7] T. Supap, C. Saiwan, R. Idem and P. P. T. Tontiwachwuthikul. 2011 Part 2: Solvent management: solvent stability and amine degradation in CO_2 capture processes *Carbon Manage.* **2(5)** 551-566.

[8] C. Saiwan, T. Supap, R. O. Idem and P. Tontiwachwuthikul 2011 Part 3: Corrosion and prevention in post-combustion CO_2 capture systems *Carbon Manage.* **2(6)** 659-675.

[9] P. Pearson, A. F. Hollenkamp and E. Meuleman 2013 Electrochemical investigation of corrosion in CO_2 capture plants—Influence of amines *Electrochim Acta* **110** 511-516.

[10] P. S. Kumar, J. A. Hogendoorn, G. F. Versteeg and P. H. M. Feron 2003 Kinetics of the reaction of CO_2 with aqueous potassium salt of taurine and glycine *AIChE J.* **49(1)** 203-213.

[11] J. Brouwer, P. Feron and N. Ten Asbroek 2005 Amino-acid salts for CO_2 capture from flue gases *Fourth Annual Conference on Carbon Capture & Sequestration*, Alexandria, Virginia, USA.

[12] P. H. M. Feron and Ten N. Asbroek. 2005 New solvents based on amino-acid salts for CO_2 capture from flue gases *Greenhouse Gas Control Technologies 7*. Oxford: Elsevier Science Ltd 1153-1158.

[13] Jv. Holst, G. F. Versteeg, D. W. F. Brilman and J. A. Hogendoorn 2009 Kinetic study of with various amino acid salts in aqueous solution *Chem. Eng. Sci.* **64(1)** 59-68.

[14] U. E. Aronu, H. F. Svendsen and K. A. Hoff. 2010 Investigation of amine amino acid salts for carbon dioxide absorption *Int. J. Greenhouse Gas Control* **4(5)** 771-775.

[15] H. Knuutila, U. E. Aronu, H. M. Kvamsdal and A. Chikukwa. 2011 Post combustion CO_2 capture with an amino acid salt *Energy Procedia* **4** 1550-1557.

[16] M. E. Majchrowicz and D. W. F. Brilman 2012 Solubility of CO_2 in aqueous potassium l-prolinate solutions—absorber conditions *Chem. Eng. Sci.* **72** 35-44.

[17] D. Kang, S. Park, H. Jo, J. Min and J. Park 2013 Solubility of CO_2 in Amino-Acid-Based Solutions of (Potassium Sarcosinate), (Potassium Alaninate + Piperazine), and (Potassium Serinate + Piperazine) *J. Chem. Eng. Data* **58(6)** 1787-1791.

[18] Y-T. Chang, R. B. Leron and Li M-H 2015 Carbon dioxide solubility in aqueous potassium salt solutions of L-proline and DL-α-aminobutyric acid at high pressures *J. Chem. Thermodyn.* **83** 110-116.

[19] S. Garg, A. M. Shariff, M. S. Shaikh, B. Lal, A. Aftab and N. Faiqa 2016 Selected physical properties of aqueous potassium salt of l-phenylalanine as a solvent for CO_2 capture *Chem. Eng. Res. Des.* **113** 169-181.

[20] S. Garg, A. M. Shariff, M. S. Shaikh, B. Lal, A. Aftab and N. Faiqa 2016 VLE of CO_2 in aqueous potassium salt of L-phenylalanine: Experimental data and modeling using modified Kent-Eisenberg model *J. Nat. Gas Sci. Eng.* **34** 864-872.

[21] S. Garg, A. M. Shariff, M. S. Shaikh, B. Lal, H. Suleman and N. Faiqa 2017 Experimental data, thermodynamic and neural network modeling of CO_2 solubility in aqueous sodium salt of l-phenylalanine *J. CO_2 Util.* **19** 146-156.

[22] S. Shen, Yang Y-n, Y. Wang, S. Ren, J. Han and A. Chen 2015 CO_2 absorption into aqueous potassium salts of lysine and proline: Density, viscosity and solubility of CO_2 *Fluid Phase Equilib.* **399** 40-49.

[23] I. Eide-Haugmo, O. G. Brakstad, K. A. Hoff, K. R. Sørheim, da Silva EF and Svendsen HF 2009 Environmental impact of amines *Energy Procedia* **1(1)** 1297-1304.

[24] B. Fischer 2013 Investigation of the Reactions when Applying Amino Acid Salts for Post-Combustion Carbon Capture Processes: Degradation Pathways, Environmental and Economic Impact *PhD thesis* University of Duisburg-Essen, Germany.

[25] M. Rabensteiner, G. Kinger, M. Koller, G. Gronald, S. Unterberger and C. Hochenauer. 2014 Investigation of the suitability of aqueous sodium glycinate as a solvent for post combustion carbon dioxide capture on the basis of pilot plant studies and screening methods *Int. J. Greenhouse Gas Control* **29** 1-15.

[26] A. L. Kohl and F. C. Riesenfeld 1985 *Gas purification*: Gulf Pub. Co.

[27] D. J. Heldebrant, P. K. Koech, J. E. Rainbolt, F. Zheng, T. Smurthwaite, C. J. Freeman, M. Oss and I. Leito. 2011 Performance of single-component CO_2-binding organic liquids (CO_2BOLs) for post combustion CO_2 capture *Chem. Eng. J.* **171(3)** 794-800.

[28] F. Barzagli, F. Mani and M. Peruzzini. 2013 Efficient CO_2 absorption and low temperature desorption with non-aqueous solvents based on 2-amino-2-methyl-1-propanol (AMP) *Int. J. Greenhouse Gas Control* **16** 217-223.

[29] V. Barbarossa, F. Barzagli, F. Mani, S. Lai, P. Stoppioni and G. Vanga 2013 Efficient CO_2 capture by non-aqueous 2-amino-2-methyl-1-propanol (AMP) and low temperature solvent regeneration *RSC Adv.* **3(30)** 12349-12355.

[30] S. Shen, Y. Bian and Y. Zhao 2017 Energy-efficient CO_2 capture using potassium prolinate/ethanol solution as a phase-changing absorbent Int. *J. Greenhouse Gas Control* **56** 1-11.

[31] N. F. A. Aziz, A. M. Shariff, M. S. Shaikh, L. K. Keong, S. Garg and A. Aftab 2016 Physical Properties of Aqueous Sodium Salt Solution of α-Methylalanine (Na-AMALA) *Procedia Eng.* **148** 444-450.

[32] A. Hartono, E. O. Mba and H. F. Svendsen 2014 Physical Properties of Partially CO_2 Loaded Aqueous Monoethanolamine (MEA) *J. Chem. Eng. Data* **59(6)** 1808-1816.

[33] S. Lee, S. I. Choi, S. Maken, H. J. Song, H. C. Shin, J. W. Park, K. R. Jang and J. H. Kim 2005 Physical Properties of Aqueous Sodium Glycinate Solution as an Absorbent for Carbon Dioxide Removal *J. Chem. Eng. Data* **50(5)** 1773-1776.

[34] M. M. Akbar and T. Murugesan 2012 Thermophysical properties for the binary mixtures of 1-hexyl-3-methylimidazolium bis(trifluoromethylsulfonyl)imide [hmim][Tf_2N] + N-methyldiethanolamine (MDEA) at temperatures (303.15 to 323.15) K *J. Mol. Liq.* **169** 95-101.

[35] L. A. Tirona, R. B. Leron, A. N. Soriano and M. H. Li 2014 Densities, viscosities, refractive indices, and electrical conductivities of aqueous alkali salts of α-alanine *J. Chem. Thermodyn.* **77.** 116-122.

[36] H. Kierzkowska-Pawlak, A. Chacuk and M. Siemieniec. 2014 Reaction kinetics of CO_2 in aqueous 2-(2-aminoethylamino)ethanol solutions using a stirred cell reactor *Int. J. Greenhouse Gas Control* **24** 106-114.

[37] Garg S, Shariff AM, Shaikh MS, Lal B, Aftab A and Faiqa N 2016 Measurement and prediction of physical properties of aqueous sodium salt of L-phenylalanine *J. Serb. Chem. Soc.* **82(7-8)** 905-919.

[38] S. Garg, G. Murshid, F. S. Mjalli and W. Ahmad 2017 Physical properties of aqueous blend of diethanolamine and sarcosine: experimental and correlation study *Chem Pap.* **71 (10)** 1799-1807.

[39] A. M. Shariff, M. S. Shaikh, M. A. Bustam, S. Garg, N. Faiqa and A. Aftab 2016 High-pressure Solubility of Carbon Dioxide in Aqueous Sodium L-Prolinate Solution *Procedia Eng.* **148** 580-587.

[40] S. Kadiwala, A.V. Rayer and A. Henni 2010 High pressure solubility of carbon dioxide (CO_2) in aqueous piperazine solutions *Fluid Phase Equilib.* **292 (1–2)** 20-28.

[41] S. N. Khan, S. M. Hailegiorgis, Z. Man, A. M. Shariff and S. Garg 2017 Thermophysical properties of concentrated aqueous solution of N-methyldiethanolamine (MDEA), piperazine (PZ), and ionic liquids hybrid solvent for CO_2 capture *J. Mol. Liq.* **229** 221-229.

[42] Shaikh MS, Shariff AM, Garg S and Bustam MA 2017 Physical properties of aqueous solutions of potassium l-prolinate from 298.15 to 343.15 K at atmospheric pressure *Chem Pap.* **71(7)** 1185-1194.

[43] Garg S, Murshid G, Mjalli FS, Ali A and Ahmad W 2017 Experimental and correlation study of selected physical properties of aqueous blends of potassium sarcosinate and 2-piperidineethanol as a solvent for CO_2 capture Chem. Eng. Res. Des. **118** 121-130.

IOP Conf. Series: Earth and Environmental Science **154** (2018) 012021 doi:10.1088/1755-1315/154/1/012021

Pyrolysis of polyethylene terephthalate containing real waste plastics using Ni loaded zeolite catalysts

M. Al-asadi[1] and N. Miskolczi[1]

1 University of Pannonia, Faculty of Engineering, Institute of Chemical Engineering and Process Engineering, MOL Department of Hydrocarbon & Coal Processing, H-8200, Veszprém, Egyetem u. 10, Hungary

E-mail: moh_iraq85@yahoo.com

Abstract. In this work the pyrolysis of polyethylene terephthalate (PET) containing real waste plastic was investigated using different Ni loaded catalysts: Ni/ZSM-5, Ni/y-zeolite, Ni/β-zeolite and Ni/natural zeolite (clinoptilolite). Raw materials were pyrolyzed in a horizontal tubular reactor between 600 and 900°C using 10% of catalysts. It was found, that both temperature increasing and catalysts presence can increase the gas yields, however owing to gasification reactions, the pyrolysis oil yield decreased with increasing temperature. Ni/y-zeolite catalyst had the most benefit in gas yield increasing at low temperature; however Ni/ZSM-5 showed advanced property in gas yield increasing at high temperature. Gases contained hydrogen, carbon oxides and hydrocarbons, which composition was significantly affected by catalysts. Ni loaded zeolites favoured to the formation of hydrogen and branched hydrocarbons; furthermore the concentrations of both CO and CO_2 were also increased as function of elevated temperature. That phenomenon was attributed to the further decomposition of PET, especially to the side chain scission reactions. Owing to the Boudouard reaction, the ratio of CO_2/CO can increased with temperature. Pyrolysis oils were the mixtures of n-saturated, n-unsaturated, branched, oxygen free aromatics and oxygenated hydrocarbons. Temperature increasing has a significant effect to the aromatization and isomerization reactions, while the catalysts can efficiently decreased the concentration of oxygen containing compounds.

1. Introduction

The environmental concern over waste plastic, especially polyethylene terephthalate (PET) and polyethylene (PE) is growing around the World. Regarding the plastic consumption, the PE and PET represent approximately 30% and 10% of the total market, respectively [1]. Mobility and fast moving people causes the increasing demand of the PE and PET for different purposes such as drinking water bottle, agriculture, constructional materials or even food packaging. On the other hand, serious environmental problems can be caused due to the dumping of these plastics in the landfill sites. Owing to the costumer habits, selectively collected waste bins and waste islands containing vast amount of PE and PET. These plastics have chemical resistance and they are non-biodegradable. Therefore, they will accumulate in the landfill sites forever if no recycling process is carried out [2], [3]. It is also known, that the high efficient selective collection has high investment and labour cost. An average PET bottle contains approximately 10-15% polyethylene, such as cup, printed label, etc. and their type selective separation is difficult.

Nowadays, among the different methods used for waste plastic recycling, pyrolysis is considered as one of the environmentally friendly techniques for waste plastic conversion into valuable products [2],

[4]. During pyrolysis the long carbon chains of the polymers are cracked at elevated temperature (400-1000°C) resulted in more valuable products, such as hydrocarbon gases, fuels, oils, chemicals, etc.
The products of pyrolysis process are affected by many factors such as, type of catalyst and its amount [4], [5]. E.g. both the product yields and composition are significantly affected by different catalyst, such as synthetic or natural zeolites, alumina, silica-alumina, etc.. In high temperature pyrolysis process, catalysts play also important role to the gases [4]-[6]. In general, higher yields of gases can be found using catalysts [5], [6]. However, many catalysts have been used for the thermal decomposition of waste (PET and PE) in order to enhance and improve the waste plastics recycling and product selectivity. Wong et al. indicated that the pyrolysis product selectivity and coke formation are affected by the pore structure and acidity of the catalyst [7]. ZSM-5, y-zeolite and other synthetic zeolite structures are widely investigated catalysts for promoting the decomposition reactions and affecting the product structure. Especially the aromatization and isomerization reactions can be enhanced and further cracking of aromatic derivate can be occurred over ZSM-5 and y-zeolite [8]-[10]. Yuan et al. found that the aromatics yield was increased by pyrolysis of PET and PE due to the hydrogen transfer from polyethylene to polyethylene terephthalate while, the yield of solid residue was decreased for the same reason [5]. Hydrogen and other valuable products can be synthetized if transition metals are built into the structure of SiO_4 and AlO_4- tetrahedral, which linked via common oxygen atoms containing ion-exchangeable mobile alkali atoms [10]-[12]. E.g. Ni, Fe or Zn resulted high hydrogen content in gases and high ratio of filamentous carbons in coked deposition on catalyst surface [12]-[15].
The goal of this work was the investigation of the pyrolysis of 50%/50% polyethylene and polyethylene terephthalate waste plastics at high temperature in tubular furnace using different Ni loaded zeolite catalyst, focussing to the effect of ZSM-5, y-zeolite, β-zeolite and natural zeolite supporters on the pyrolysis process. Especially the product yields and the compositions of gaseous and pyrolysis oil fraction were discussed.

2. Experimental

2.1. Raw material

In this work scraps of real plastic waste were used: polyethylene terephthalate was obtained from waste mineral water bottles, while polyethylene wastes was obtained from packaging. Both raw materials were shredded and milled by laboratory grinder (Dipre GRS 183A9) into small particles (<5mm). The main properties of raw materials are listed in Table 1.

Table 1. Main properties of raw materials.

	Polyethylene	**Polyethylene terephthalate**
Polymer	$-(CH_2-CH_2)_n-$	$-(O-CO-C_6H_4-CO-O-CH_2-CH_2)_n-$
Source	packaging	soft drink/mineral water bottles
Ash content [a], **%**	2.5	4.8
Humidity, %	1.02	1.44

[a] combustion in air at 850°C for 2hours

2.2. Catalysts

Mixtures of raw materials were pyrolyzed over different nickel loaded zeolite supported catalysts. They were synthetized using ZSM-5, y-zeolite, β-zeolite and natural zeolite (clinoptilolite) catalysts according to the following method. Catalyst supporters were added into 1M $Ni(NO_3)_2 \cdot 6H_2O$ dissolution and continuously stirred for 2 hours at 85°C. Then catalysts were dried for 10 hours at 110°C, and each of them was conditioned at 650°C for 4.5 hours in air. According to x-ray analysis, Ni/ZSM-5, Ni/y-zeolite, Ni/β-zeolite and Ni/natural zeolite catalysts had nickel content of 10.4, 10.1, 10.3 and 10.7%, while they had 315, 610, 580, 205m^2/g BET surface area, respectively.

IOP Conf. Series: Earth and Environmental Science **154** (2018) 012021 doi:10.1088/1755-1315/154/1/012021

2.3. Pyrolysis apparatus

Mixtures of 50% waste polyethylene and 50% polyethylene terephthalate were pyrolyzed in a tubular reactor in nitrogen flow at 600, 750 and 900°C. Figure 1 demonstrates the layout of experimental rig. 10g of raw materials were weighted into the sample holder. In case of thermo-catalytic pyrolysis 10% of the synthetized catalysts were also added to the raw materials. Then nitrogen was continuously flushed through a quartz reactor tube during the experiments. Pyrolysis was taken at 600, 750 and 900°C till 20 minutes. The reactor temperature was set and controlled by PID controller. Pyrolysis products were driven through a cooled glass container, where liquid fraction was condensed. Gases were taken from the container and further analysed. The temperature of the cooling water was 5°C.

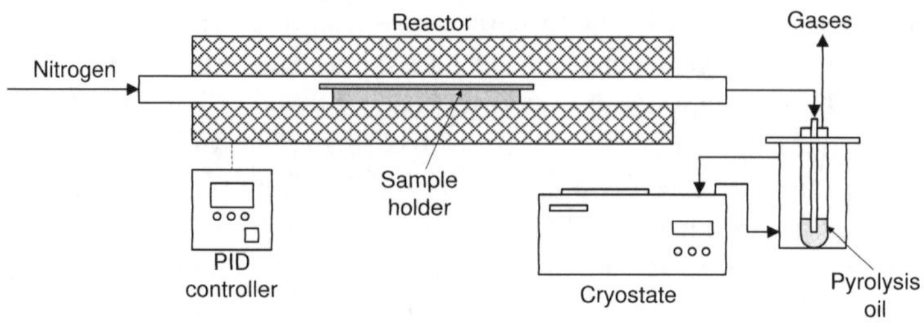

Figure 1. Layout of the pyrolysis process

2.4. Product analysis

The composition of gases was investigated by a DANI type GC instrument, which was fitted with programmed injector, flame ionized detector and Rtx PONA (100m x 0.25mm, surface thickness of 0.5 μm) and Rtx-5 PONA (100m x 0.25mm, surface thickness of 1 μm) columns. Sample analysis was taken using isotherm conditions (T=30°C).

Hydrogen content of the gas products was measured by Shimadzu GC-2010 gas-chromatograph equipped with TCD detector (CarboxenTM 1006 PLOT column (30m × 0.53mm)). The temperature program was the following: from 35°C (hold time 2 min) to 250°C at 40°C min^{-1} heating rate, with the final temperature maintained for 5 min. The GC instrument was previously calibre by mixture of defined composition of gases.

Pyrolysis oil was analysed by DANI GC, using Rtx 1 dimetil-polysiloxan capillary column (30m x 0.53mm, thickens of 0.25 μm). Sample was heated to 40°C for 5 minutes, then the temperature was elevated by 8°C/min till 340°C and it was kept at 340°C till 20min. Both the injector and detector temperature was 340°C.

A TENSOR 27 type Fourier transformed infrared spectrometer (resolution: 2 cm^{-1}, illumination: SiC Globar light, monochromator: Littrow prism, detector: RT-DLaTGS (Deutero L-α-Alanine Triglycine-Sulphate) type detector with KBr window) was also used for pyrolysis oil analysis in wavenumber range of 400-4000cm^{-1}.

3. Result and discussion

3.1. Product yields

The yields of gases and pyrolysis oil obtained in the decomposition reactions are summarized in Figures 2 and 3. Results indicate that the yields of both gases and pyrolysis oil were much higher by the using of catalysts compared with without catalyst. It is well known, that pyrolysis needs lots of energy, which could be decreased by catalysts. E.g. the activation energy and the reaction time could be significantly decrease by catalysts, which led to decreasing in energy consumption [3]. For the pyrolysis process without catalyst, the results were 21.4, 35.2 and 48.3 for temperature of 600, 750 and 900°C, respectively. It is well known, that the thermal stability of C-C bond decreases with increasing temperature. That is the cause for, that higher pyrolysis temperature can promote the more intensive cracking of C-C bond, leading to the dominant formation of gases. Figure 2 well illustrates that catalysts can also promote the more intensive cracking of C-C bonds, leading to significantly increase in the yield of gaseous products in the temperature range of 600-900°C. It can be seen that the gases yields over Ni/ZSM-5 catalyst was the highest when compared to the other three catalysts. Ni/ZSM-5 catalyst increases the gas yields with 42.4-69.2% depending on the reaction temperature. Natural zeolite supported Ni modified catalysts had the less effect to the decomposition reactions, because that catalysts can increase the gas yield only with 5.8-27.6%. It is important to remark, that the catalysts efficiency was different at 600 and 900°C, because it was the order of Ni/Natural zeolite<Ni/β-zeolite<Ni/ZSM-5<Ni/y-zeolite at 600°C, however Ni/Natural zeolite<Ni/y-zeolite<Ni/β-zeolite<Ni/ZSM-5 at 900°C. The high cracking activity could be attributed to the high surface area of Ni/β-zeolite, Ni/y-zeolite and Ni/ZSM-5 catalysts; Ni/natural zeolite catalysts had lower BET surface area.

The yields of pyrolysis oil for the four catalysts are shown in Figure 3, which were decreased by increasing temperature. The pyrolysis oil yields were 34.4, 21.6 and 11.4% at 600, 750 and 900°C without catalyst, respectively. Ni/ZSM-5 catalyst had the highest activity in pyrolysis oil yield increasing at 600°C with yield of 52.9%. β-zeolite showed also high activity in pyrolysis reactions resulted pyrolysis oils, however its activity rapidly decreased with increasing temperature. It was only 19.3% at 900°C. Glancing result, it was concluded, that Ni/y-zeolite catalyst had the slowest decreasing tendency. Comparing gas and pyrolysis oil yields, it is also clear, that basically the transformation of pyrolysis oil to gaseous product is the main reason to the decreasing tendency in pyrolysis oil yields as function of temperature. Higher temperature favours to the deeper cracking of primary decomposition products.

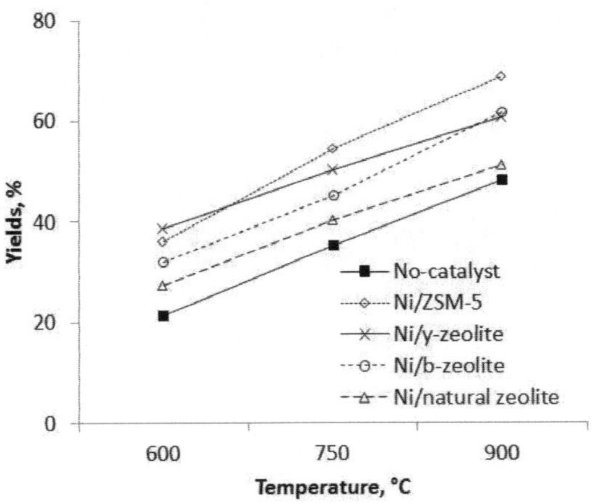

Figure 2. Yields of gases as function of temperature

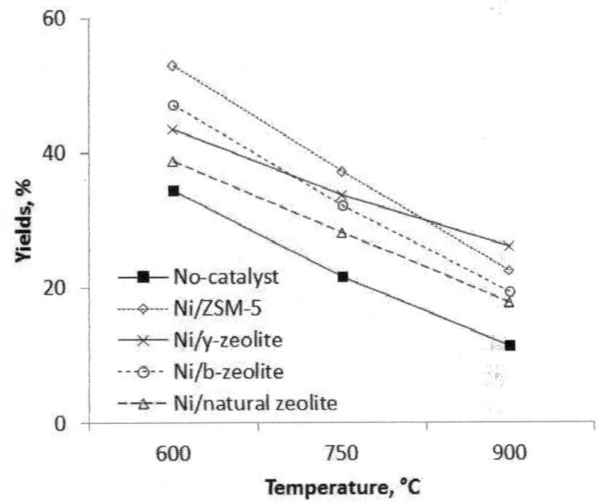

Figure 3. Yields of pyrolysis oil as function of temperature.

3.2. Gases

The compositions of gases are summarized in Figure 4, which shows the effect of catalyst and reaction temperature to the pyrolysis process with catalysts and without catalyst. Gases contain hydrogen, CO, CO_2, branched and non-branched hydrocarbons up to C_5. Regarding the effect of the pyrolysis temperature, it is well shown that the hydrogen yields could increase with increasing temperature, both absence and presence of catalysts. E.g. it was 4.0% at 600°C and 14.1% at 900°C without catalyst. Hydrocarbons resulted hydrogen and coke at high temperature as following reaction:

$$C_nH_m \rightarrow C_kH_l + C + H_2$$

Higher temperature favours to the carbon and hydrogen production. Higher temperature favours to intramolecular hydrogen transfer reaction, which resulted higher yields of branched C_2-C_5 hydrocarbons with increasing temperature. CO and CO_2 in gases were the consequence of PET presence in raw materials. PET contains -(O-CO-C_6H_4-CO-O-CH_2-CH_2)- repeating unit in the polymer macromolecule, which decomposition results significant carbon oxides. In general CO_2 had higher concentration than CO. In the most cases, the ratio of CO_2/CO increased with temperature, demonstrating the higher rate of Boudouard reaction at higher temperature. However, owing to PET structure, the water production and in-situ water-gas shift reaction or steam-methane reforming reaction in the produced water presence should be also the cause for CO_2 increasing. Regarding CO increasing the water-gas reaction could be also blamed.

$$\mathrm{CO} + H_2O \rightarrow \mathrm{CO_2} + H_2$$
$$\mathrm{CH_4} + 2H_2O \rightarrow \mathrm{CO_2} + 4H_2$$
$$\mathrm{C} + H_2O \rightarrow CO + H_2$$
$$\mathrm{CO_2} + C \rightarrow CO$$

The concentration of methane did not changed significantly with temperature, however it was highest a bit using higher process temperature. The increasing in methane concentration should be occurred by the methanization reaction: $C+2H_2 \rightarrow CH_4$.

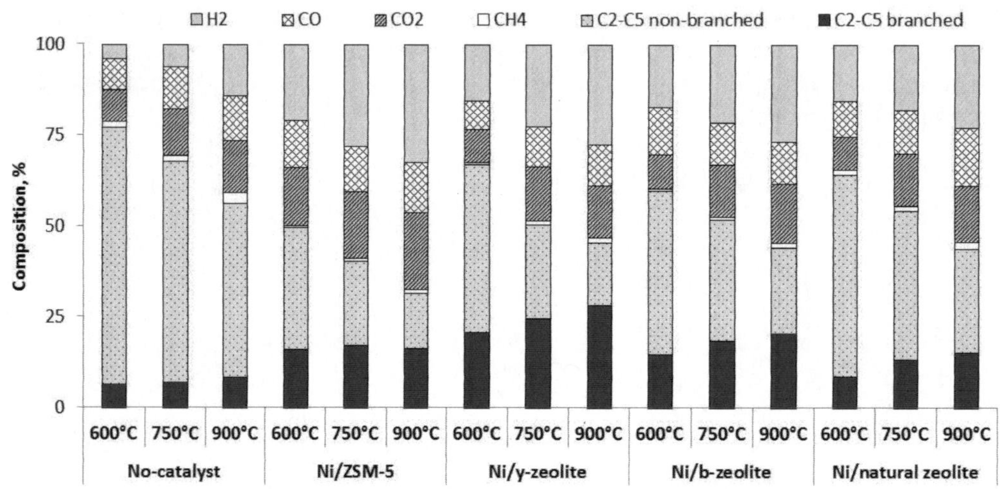

Figure 4. The composition of gases

Catalysts had also important effect to the composition of gaseous fractions. E.g. they can increase the yields of hydrogen and branched hydrocarbons and decreased the concentration of non-branched hydrocarbons. The increase in branched hydrocarbon was attributed to the intramolecular and intermolecular hydrogen transfer reactions over acidic zeolite catalyst supporters. Regarding isomerization, Ni/y-zeolite had the highest activity; the concentrations of branched hydrocarbons were 20.8, 24.5 and 28.1% at 600, 750 and 900°C, respectively. However, Ni modified natural zeolite showed the less effect in branching reactions. Both CO and CO_2 had higher concentrations by the using of catalysts, which was the consequence of more intensive decomposition of PET, lollowed by the before listed secondary reactions. The most important effect of the catalyst was the increasing of hydrogen content. E.g. gases synthesised over Ni/ZSM-5 catalyst had 32.6% hydrogen at 900°C. Presumably transition metal loaded zeolite can effectively increase the rate of reaction led to hydrogen formation; e.g. water-gas reaction, water-gas shift reaction or even decomposition of main carbon frame of polyethylene resulted coke, light hydrocarbons and hydrogen.

3.3. Pyrolysis oil

The compositions of pyrolysis oils are shown in Figure 5. Results indicate that the oil product contained n-olefins, n-paraffins, non-oxygenated aromatics, branched and oxygenated compounds. The main effect of the temperature increasing was the increasing in non-oxygenated aromatics and decreasing in oxygenated compounds. The concentrations of n-alkane, n-alkene did not change significantly with temperature, however lower yields of n-olefins and n-paraffins were measured in pyrolysis oils obtained in thermo catalytic pyrolysis. N-olefins and n-paraffins were represented with concentrations of 31.9-33.3% and 23.3-24.3% without catalysts, while 20.2-27.0% and 15.4-21.7% in thermo-catalytic pyrolysis.

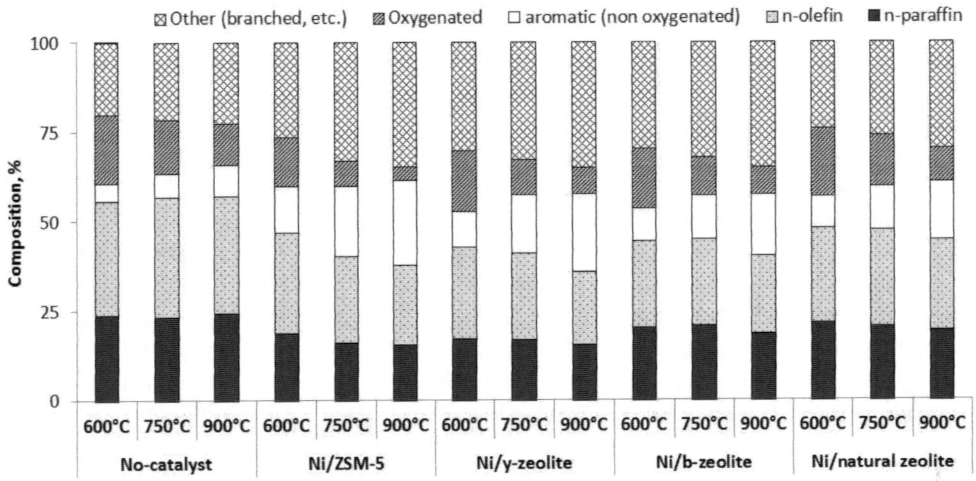

Figure 5. The composition of pyrolysis oils

Based on FTIR result, oxygenated compounds were represented mainly in the form of carboxylic acids, esters, phenols and its derivates according to the following reaction scheme:

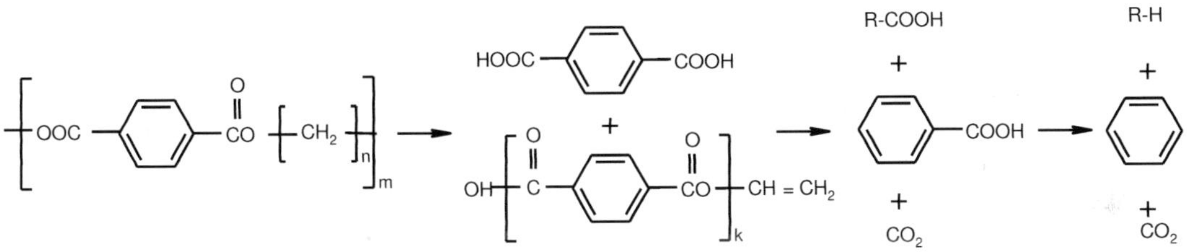

It is also clear, that the temperature increases and catalyst presence favour to the further decomposition of carboxylic acids, such as terephtalic acid, acetic acid, benzoic acid, or oligomers of PET, resulting oxygen free aromatics and CO_2. Aromatization reactions and side broken reactions of phenol derivates (especially from PET pyrolysis) could be also increased with temperature. Both the elevated temperature and presence of acidic catalysts are favoured for them.

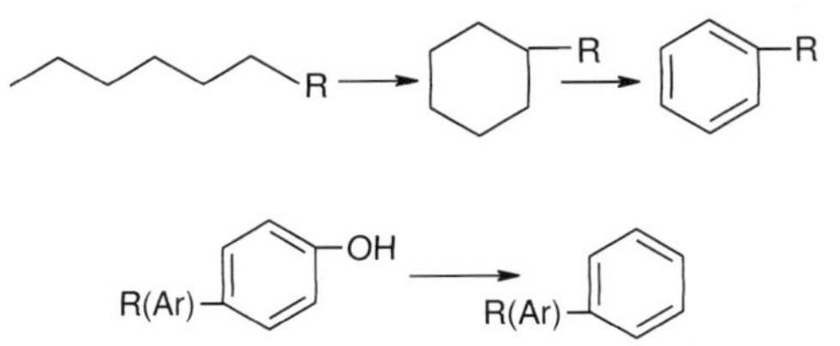

As it was before mentioned, higher pyrolysis temperature favours to the intramolecular hydrogen transfer reactions, which lead to isomerisation reactions. That is why that higher concentration of branched hydrocarbons was obtained e.g. at 900°C.

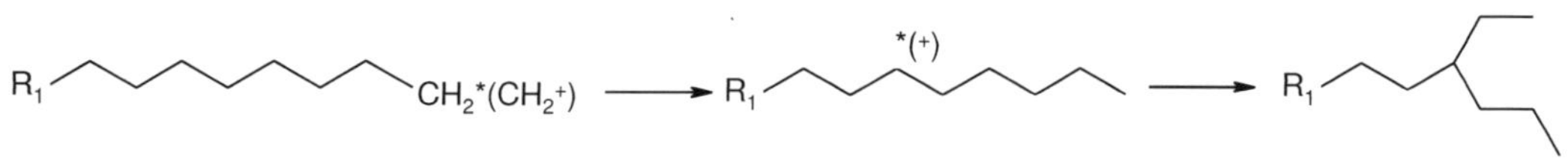

In pyrolysis reaction using catalysts, higher yield of branched compounds was achieved over Ni/y-zeolite and Ni/β-zeolite, while the maximum yields of aromatic compounds was achieved over Ni/ZSM-5 which is.

4. Summary
In this work, the co-pyrolysis of waste polyethylene and polyethylene terephtalate was investigated using different Ni modified zeolite catalysts. Both gas and pyrolysis oily yields were significantly influenced by temperature and catalysts: temperature could increase the gas and decrease the pyrolysis oil yield. In gases H_2, CO_2, CO, CH_4, branched and non-branched hydrocarbons up to C_5, while in pyrolysis oil branched, oxygenated, aromatic, n-olefin and n-paraffin were mainly represented. Ni/ZSM-5 can effectively increase the amount of hydrogen. The maximum hydrogen production was achieved at high temperature therefore the high temperature is favoured for hydrogenation reaction. On the other hand the maximum oil product was achieved over Ni/y-zeolite and Ni/β-zeolite. Owing to the side chain scission and secondary reactions, both catalysts and temperature increasing decreased the concentrations of oxygenated aromatic and increased the concentration non-oxygenated compounds or even CO_2.

5. Acknowledgments
The authors acknowledge the National Office for Research and Innovation (NKTH), assistance granted under the Hungarian-Indian (KTIA-DST) R&D&I Program (TÉT_13_DST-1-2014-0003). We also acknowledge the financial support of Széchenyi 2020 under the EFOP-3.6.1-16-2016-00015.

6. References
[1] http://www.plasticseurope.org/Document/plastics---the-facts-2016-15787.aspx?FolID=2
[2] Z. Sun, B. Xu, A. H. Rony, S. Toan, K. A. M. Gasem, H. Adidharma, M. Fan, M. Xiang, *Energ Convers Manage,* 2017. **146** 182-194
[3] N. Miskolczi, R. Nagy, *Fuel Proc Technol,* 2012. **104** 96-104
[4] S. M. Al-Salem, A. Antelava, A. Constantinou, G. Manos, A. Dutta, *J Environ Manage,* 2017. **197** 177-198
[5] X. Yuan, P. Johnston, X. Bai, *Energ Convers Manage,* 2017. **142** 441-451.
[6] Miandad R, Barakat MA, Rehan M, Aburiazaiza AS, Ismail IMI, Nizami AS, *Waste Manage,* 2017. **69** 66-78
[7] H. W. Wong, J. Peck, J. Assif, F. Panerai, F. Lachaud, N. N. Mansour, *J Anal Appl Pyrol,* 2016. **122** 258-267
[8] S. Kumagai, R. Yamasakic, T. Kameda, Y. Saito, A. Watanabe, C. Watanabe, N. Teramae, T. Yoshioka. *Chem Eng J,* 2018. **332** 169-173
[9] J. Chattopadhyay, T. S. Pathak, R. Srivastava, A. C. Singh, *Energy,* 2016, 103, 13-521
[10] M. Artetxe, G. Lopez, M. Amutio, G. Elordi, J. Bilbao, M. Plazar, *Ind Eng Chem Res,* 2013. **52**, 10637-10645.
[11] A. R. Auxilio, W. L. Choo, I. Kohli, S. C. Srivatsa, S. Bhattacharya, *Waste Manage,* 2017. **67** 143-154
[12] A. Ochoa, I. Barbarias, M. Artetxe, A. G. Gayubo, M. Olazar, J. Bilbao, P. Castaño, *Appl Catal B- Environ,* 2017. **209** 554-565
[13] J. C. Acomb, C. Wu, P. T. Williams, *J Anal Appl Pyrol,* 2015. **113** 231-238
[14] D. Yao, C. Wu, H. Yang, Y. Zhang, A. M. Nahil, Y. Chen, P. T. Williams, H. Chen, *Energ Convers Manage,* 2017. **148** 692-700
[15] E. F. Iliopoulou, S. D. Stefanidis, K. G. Kalogiannis, A. Delimitis, A. A. Lappas, K. S. Triantafyllidis, *Appl Catal B- Environ,* 2012. **127** 281-290

Co-pyrolysis of biomass and plastic wastes: investigation of apparent kinetic parameters and stability of pyrolysis oils

B. Fekhar[1], N. Miskolczi[1], T. Bhaskar[2], J. Kumar[2] and V. Dhyani[2]

[1] University of Pannonia, Faculty of Engineering, Institute of Chemical Engineering and Process Engineering, MOL Department of Hydrocarbon & Coal Processing, H-8200, Veszprém, Egyetem u. 10, Hungary
[2] CSIR-Indian Institute of Petroleum, Thermo-catalytic Processes Area, Dehradun 248005, India

Abstract. This work is dedicated to the co-pyrolysis of real waste high density polyethylene (HDPE) and biomass (rice straw) obtained from agriculture. Mixtures of raw materials were pyrolyzed in their 0%/100%, 30%/70%, 50%/50%, 70%/30%, 100%/0% ratios using a thermograph. The atmosphere was nitrogen, and a constant heating rate was used. Based on weight loss and DTG curves, the apparent reaction kinetic parameters (e.g., activation energy) were calculated using first-order kinetic approach and Arrhenius equation. It was found that decomposition of pure plastic has approximately 280 kJ/mol activation energy, while that of was considerably less in case of biomass. Furthermore, HDPE decomposition takes by one stage, while that of biomass was three stages. The larger amount of raw materials (100 g) were also pyrolyzed in the batch rig at 550°C to obtain products for analysis focussing to their long-term application. Pyrolysis oils were investigated by Fourier transformed infrared spectroscopy and standardized methods, such as density, viscosity, boiling range determination. It was concluded, that higher plastic ratio in raw material had the advantageous effect to the pyrolysis oil long-term application. E.g., the concentration of oxygenated compounds, such as aldehydes, ketones, carboxylic acids or even phenol and its derivate could be significantly decreased, which had an advantageous effect to their corrosion property. Lower average molecular weight, viscosity, and density were measured as a function of plastic content.

1. Introduction

Plastics are unique and widely used materials. The worldwide plastic demand has increased rapidly over recent years, and the production reached the 300 million tons at 2015 [1]. The growth in plastic consumption is also responsible for the increase in municipal solid wastes; therefore the long-term sustainable utilization of plastic and municipal solid wastes is in the focus of waste management. The total municipal solid waste production was around 3.5 billion tonnes in 2015 at Worldwide. Plastic wastes are also responsible for significant environmental problems because they are mainly landfilled and incinerated, which are not to be the sustainable option for long-term waste plastic utilization. On the other hand, plastic waste has high "hidden" energy content. An average municipal plastic waste has a heating value of ~25 MJ/kg, while that of ~15 MJ/kg in case of biomass depending on the composition and water content. It means that significant energy and emissions of greenhouse gases could be saved by their proper recycling. Owing to the extremely slow degradation of plastics, the landfilling needs huge lands, while incinerators have significant exhaust gas emission and needs expensive gas filter unite in the chimney. It is also known, that different plastics should not be blended during the mechanical recycling processes, therefore the selective collection and separation is the weakest point in that recycling. According to report, with less landfill, average incineration, mainly chemical recycling 70-75MJ/kg waste energy could be saved, while with incineration with energy recovery only 5-10MJ/kg. Regarding CO_2 emission reduction, also the less landfill, average incineration, mainly chemical recycling showed the best option.

The pyrolysis as a process for waste polymer (plastic and biomass) chemical recycling needs high temperatures (400-1000°C) because the raw materials (especially plastics) have low thermal conductivity [2]. The product yields and properties are significantly depended by the reaction parameters; e.g., temperatures, catalysts, raw materials, etc. High pyrolysis temperature favours to increasing in volatile yields, while the catalyst can support the thermal degradation reactions at low temperatures with lower energy consumption [2]-[6]. On the other hand, the cracking reactions are faster leading to less reaction/residence time, furthermore reactor capacity should be smaller using catalysts and higher temperatures [4]. Another important parameter is the raw material because it has a huge impact to the product properties. Non-contaminated hydrocarbons are the products in case of polyethylene, polypropylene and polystyrene pyrolysis, while oxygenated hydrocarbons (aldehydes, ketons, carboxylic acids, phenol and derivates, etc.) could be obtained by biomass pyrolysis [7], [8]. The long-term application of pyrolysis oils are fundamentally affected by contaminates. Chlorine, nitrogen, aldehydes, ketones, carboxylic acids, phenol, and derivates can cause corrosion problems, while the polymerization of unsaturated hydrocarbons leads to storage and transportation problems. Therefore the reductions of before listed components are key for pyrolysis oil long-term application.

It is also well known that the mechanism of thermal degradation of waste plastic is very complex including the following reactions: initiation, propagation, and termination. To understanding the characteristic of waste polymer decomposition it is essential to investigate the reactions leading to volatile products. Numerous papers on waste polymer pyrolysis kinetics have been performed in literature using different reaction kinetic approaches. It can be concluded, that the decomposition reactions are modelled basically as first-order kinetic models and the thermogavimetric method is widely used for analysis connected with coupled techniques, e.g., TG-MS, pyrolysis-TG-MS, etc. [9], [10].

The goal of this work is the investigation of reaction kinetic of real waste HDPE and biomass co-pyrolysis and calculations of apparent reaction kinetic parameters. Pyrolysis was also performed in batch reactor to obtain larger volume of pyrolysis oils, and then the products were characterized focussing to the long-term application regarding their corrosion and transportation properties.

2. Experimental

2.1. Raw material

Mixtures of real plastic waste high density polyethylene (HDPE) and biomass wastes (rice straw) were used as raw material. Both of them were obtained from agriculture sector. Raw materials were milled in a laboratory grinder (Dipre GRS 183A9) into particles up to 0.5 mm. Raw materials were used in their 100%/0%, 70%/30%, 50%/50%, 30%/70% and 0%/100% weight ratio. Table 1 summarizes the main properties of the raw materials.

Table 1. Main properties of raw materials.

	Polyethylene	**Biomass**
Polymer	LDPE, HDPE	Rice husk
Source	agriculture	agriculture
Ash content [a], %	4.1	8.8
Humidity, %	1.02	7.61

[a] combustion in air at 850°C for 2hours

2.2. TG instrument and calculation

The apparent kinetic parameters of the decomposition reaction and the effects of the raw materials on the pyrolysis were investigated using a MOM DERIVATOGRAPH Q-1500 D type instrument. 200

IOP Conf. Series: Earth and Environmental Science **154** (2018) 012022 doi:10.1088/1755-1315/154/1/012022

mg of raw material was weighted in the sample holder and placed in the instrument. Then 20°C/min heating rate and 40 ml/min nitrogen flow was used for analysis.

The apparent activation energy of the decomposition and pre-exponential parameter were calculated using the Arrhenius equations and first-order kinetic approach:

$$-\frac{dx}{dt} = k \cdot x^n$$

$$k = A_0 \cdot e^{\left(-\frac{E}{RT}\right)}$$

where "E" is the activation energy, "A_0" is the preexponential constant, "m" is the weight of sample, "n" is the reaction order, "k" is the reaction rate coefficient, "t" is the time of degradation, "T" is the temperature and "R" is the universal gas constant. The weight loss fraction of polymer sample was calculated by:

$$x = \frac{m_i - m_t}{m_i - m_f} \cdot 100$$

where "m_i", "m_t" and "m_f" are the initial weight of sample, sample weight at "t" time and final weight of sample, respectively. The weight loss is the function of activation energy, therefore the following equation could be summarized:

$$\frac{dx}{dt} = A_0 \cdot e^{\left(-\frac{E}{RT}\right)}(1 - x)$$

2.3. Pyrolysis in batch reactor

100g of raw materials were also pyrolyzed in an electrically heated batch reactor at 550°C in nitrogen atmosphere. The reactor temperature was controlled by PID controller. Volatiles from the reactor were condensed and separated into pyrolysis oil and gases in a phase separator (Figure 1). Gases were driven through gas flow meter and a scrubber filled with 1M NaOH dissolution. Condensed fraction was further separated into water and pyrolysis oil in a separator in case of pyrolysis using biomass containing raw material.

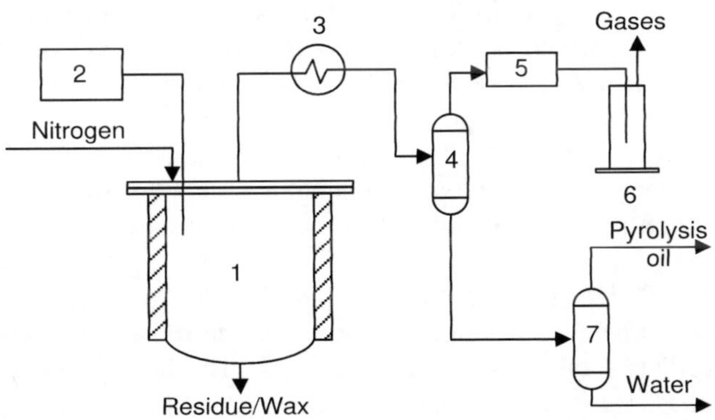

Figure 1. Batch apparatus for plastic/biomass co-pyrolysis (1-Reactor, 2-PID controller, 3-heat exchanger, 4-gas/liquid separator, 5-gas flow meter, 6-scrubber, 7-oil/water separator)

The composition of pyrolysis oil was also analyzed by TENSOR 27 type FTIR-ATR spectrometer in wave-number range of 4000-400 cm^{-1} equipped with an Attenuated Total Reflectance (ATR) accessory (Ge crystal) at room temperature. The uniform number of 32 scans with resolution 3 cm^{-1} was maintained in all case. Pyrolysis oils were also measured by standardized methods.

3. Result and discussion

3.1. Apparent reaction kinetic parameters

The weight loss and DTG curves of pyrolysis are summarized in Figures 2 and 3. Results well demonstrate that significant differences were found among decompositions of raw materials with different plastic/biomass ratio. E.g., biomass sample has three stages decomposition, while HDPE plastic waste has only one.

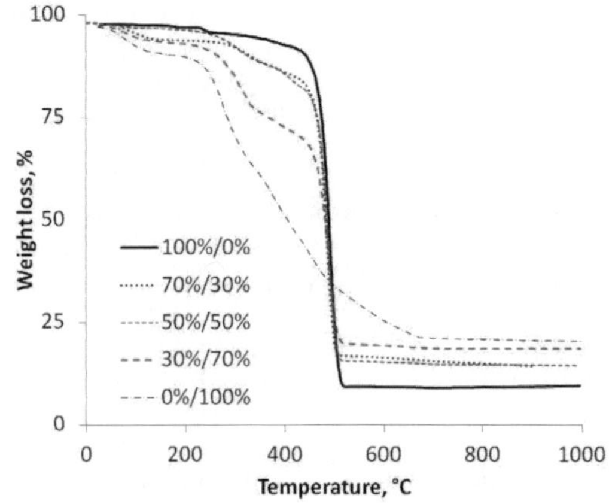

Figure 2. Weight loss of samples using different polymer/biomass ratio

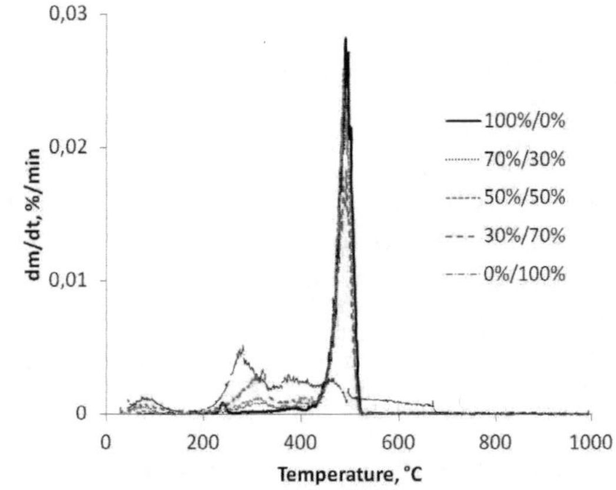

Figure 3. DTG curves of samples using different polymer/biomass ratio

In 100% biomass and 30%/70% polymer/biomass samples, the first peak around 100°C can be attributed to their physical dewatering. Biomass sample contain water, which is the reason for 7.61% humidity content. Owing to that, the conversion was between 0.5 and 7% in case of 100% biomass, while 0.5% and 4% in case of 30%/70% polymer/biomass mixture. The drying was finished till 125°C with peak maximums of 58 (biomass) and 61°C (polymer/biomass). The calculated activation energies of that stage were 26 and 28 kJ/mol, respectively.

The second stage in diagrams (1st decomposition stage) starts from 1-10% conversion depending on the raw materials and finished at 4-40% conversion. It is also clear, that the conversion range decreased with increasing ratio of HDPE in raw material. E.g., it was 1-4% using 70%/30% polymer/biomass sample or 10-40% in case of biomass. Similar trend was found regarding temperatures (T_i, T_{max} and T_f), which decreased by decreasing biomass content. E.g., the T_{max} was 275°C using 100% biomass and 262°C in 70%/30% polymer/biomass ratio. Based on the first-order reaction kinetic approach, the following apparent activation energies could be calculated: 53, 55, 58 and 63kJ/mol in case of 0%/100%, 30%/70%, 50%/50%, 70%/30% and 100%/0% raw materials, respectively. That stage is appears only biomass containing samples and missing in case of HDPE polymer. According to literature, basically the cellulose and hemicellulose are decomposed in that temperature range.

Regarding 0%/100% and 70%/30% polymer/biomass samples, the third stage (2nd decomposition stage) could be attributed to the lignin decomposition. It is important to remark, that decomposition of HDPE also occurs in that range. That stage starts around 350°C and finished till 506°C with peak maximum at 455°C using only biomass sample. It is important observation that the initial temperature increased with increasing HDPE content of the samples, while the T_{max} values were independent from concentration of polymer. On the other hand the T_f values decreased with increasing polymer/biomass ratio. Based on the calculations, it was concluded, that the conversion range was 10-90% regarding 100% HDPE waste with 272kJ/mol activation energy, which could be decreased by biomass. Regarding 70%/30%, 50%/50%, 30%70% and 0%/100% polymer/biomass raw materials, the calculated apparent activation energies were 244, 168, 136 and 62kJ/mol, respectively.

3.2. Pyrolysis oils obtained from batch reactor

Plastic and biomass wastes were also pyrolyzed in batch reactor at 550°C. The yields of volatiles and the appearance of pyrolysis oils are shown in Figures 4 and 5. Biomass pyrolysis resulted dominant amount of gases (29.5%), while plastic decomposition rather pyrolysis oil (24.2%). However the yield

IOP Conf. Series: Earth and Environmental Science **154** (2018) 012022 doi:10.1088/1755-1315/154/1/012022

of volatiles (gases+pyrolysis oil) was significantly higher in case of biomass decomposition. It is important observation that the pyrolysis oil yield was in very narrow range: 24.2-27.6%.

Table 2. Main properties of raw materials.

		Polymer/biomass ratio				
		100/0	**70/30**	**50/50**	**30/70**	**0/100**
1st peak	**Conversion, %**	-	-	-	0.5-4	0.5-7
	T_i, °C	-	-	-	50	50
	$T_{max.}$, °C	-	-	-	58	61
	T_f, °C	-	-	-	120	125
	E, kJ/mol	-	-	-	28	26
	A	-	-	-	6.36	6.23
2nd peak	**Conversion, %**	-	1-4	2-6	5-24	10-40
	T_i, °C	-	172	182	212	222
	$T_{max.}$, °C	-	262	263	288	275
	T_f, °C	-	303	307	332	339
	E, kJ/mol	-	63	58	55	53
	A	-	2.65	3.63	3.07	4.54
3rd peak	**Conversion, %**	10-90	15-90	20-95	35-95	45-85
	T_i, °C	446	450	459	475	346
	$T_{max.}$, °C	484	483	485	485	455
	T_f, °C	505	512	515	517	556
	E, kJ/mol	272	244	168	136	62
	A	28.82	25.22	20.44	8.619	10.81

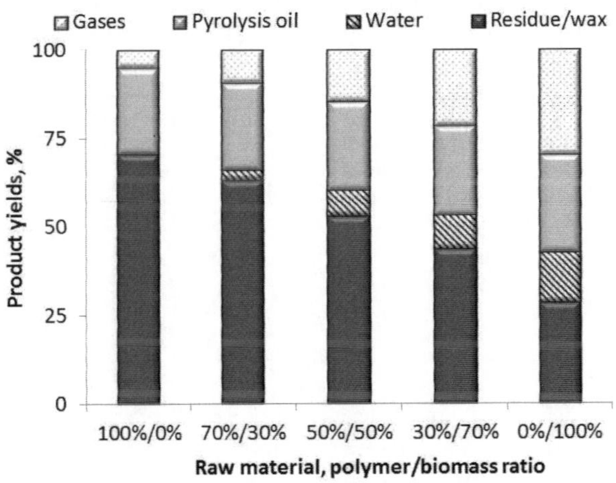

Figure 4. Product yields using different polymer/biomass ratio (t=60min.)

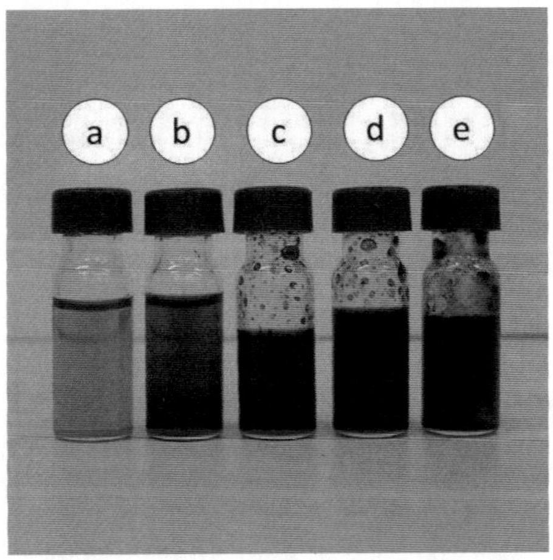

Figure 5. The appearances of pyrolysis oils (a-100%/0%, b-70%/30%, c-50%/50%, d-30%/70% and e-0%/100%)

Owing to the glycosidic hydroxyl group in cellulose structure, the water yield was also significant in biomass pyrolysis, which was significantly decreased with increasing polymer content of raw materials. Figure 5 well demonstrate that the colour of the pyrolysis oil becomes darker and less transparent with increasing biomass content of raw material.

Pyrolysis oils were characterized as fuel, focussing to the long-term storage and corrosion properties. The FTIR spectra of products are shown in Figure 6. There are well separated intensive signal between 2800 and 3000 cm^{-1}, which are attributed to the symmetric and asymmetric vibration of $-CH_2-$

(($\nu_{as}CH_2$, ν_sCH_2)) and $-CH_3$ ($\nu_{as}CH_3$, ν_sCH_3) groups. Unsaturated hydrocarbons are responsible for bands between 800 and 1000 cm^{-1}: vinyl (910 and 990 cm^{-1}), vinylidene (890 cm^{-1}) and vinylene (960 cm^{-1}). Oil fraction obtained from waste HDPE pyrolysis had intensive signal at 990 cm^{-1}, referring to the dominant $-CH=CH_2$ content, however it was decreased with increasing biomass content. Not only the $-CH=CH_2$ content, but also the concentrations of $-CH_2-$ and $-CH_3$ groups also decreased with increasing biomass/polymer ratio. That is the cause for reduced intensities of infrared signals between 2800 and 3000 cm^{-1}. Regarding vinylidene unsaturated hydrocarbons, more intensive sharp peaks was found using higher biomass content in raw materials. The high concentration of unsaturated hydrocarbons are unflavoured in case of fuel-like utilization of pyrolysis oil, because owing to polymerization of C=C bonds agglomerated macromolecules can be synthetize with less solubility and melting point. That is why they can cause storage and transportation problems. It is well known, that the pyrolysis of biomass resulted alcohols, aldehydes, ketones and carboxylic acids, which are responsible for deteriorations in storage and corrosion properties of pyrolyis oils. Infrared bands between 1140 and 1230 cm^{-1} are caused by phenolic hydrocarbons, which was the most intensive in case of biomass derived pyrolysis oils. Aldehydes (unsaturated, non-conjugated, conjugated, aromatic) resulted infrared bands in 1680-1740 cm^{-1}, however esters also shows infrared activity between 1720 and 1740 cm^{-1} wave number range.

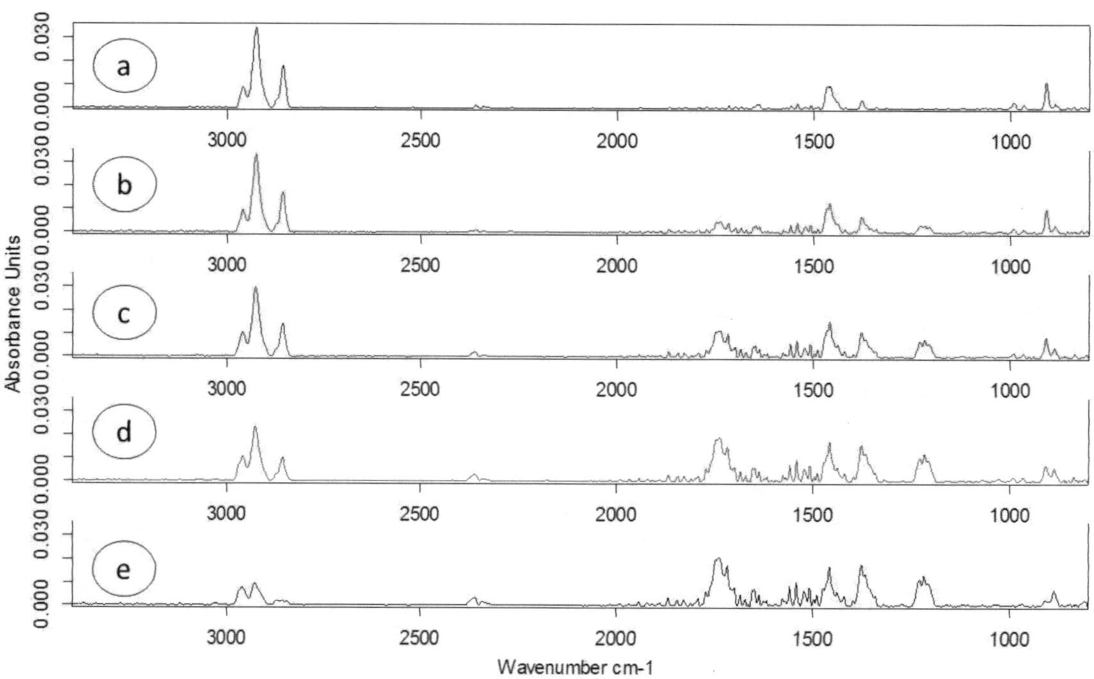

Figure 6. The FTIR spectra of pyrolysis oils
(a-100%/0%, b-70%/30%, c-50%/50%, d-30%/70% and e-0%/100%)

The main properties of pyrolysis oils are summarized in Table 3. Pyrolysis oils have densities of 0.810-0.897 g/cm^3, and kinematic viscosities of 2.1-9.7 mm^2/s. The average molecular weight was between 185 and 195 g/mol. Density, viscosity and average molecular weight are decreased with increasing plastic ratio in raw materials. The result of corrosion test show, that 100% plastic derived pyrolysis oil had excellent corrosion property, while 100% biomass derived had "bad". The quality of pyrolysis oils could be improved by increasing plastic ratio regarding corrosion aspect of plastic/biomass raw materials. Pyrolysis oils had boiling point ranges of 22-341°C, which did not changed with raw materials. Plastic derived pyrolysis oil contained saturated and unsaturated hydrocarbons; while biomass derived pyrolysis oil had significant aromatic content.

Table 3. Main properties of pyrolysis oils

	100/0	70/30	50/50	30/70	0/100
Density, g/cm3	0.810	0.828	0.852	0.866	0.897
Viscosity, mm2/s	2.1	3.8	5.5	7.1	9.7
Corrosion test	excellent	good	medium	bad	bad
M, g/mol	185	185	190	190	195
Ash content, %	<0.01	<0.01	<0.01	<0.01	<0.01
Saturated, %	48	46	39	33	29
Unsaturated, %	52	49	45	41	40
Aromatic, %	-	5	16	26	31
Boiling range, °C	25-334	23-335	24-340	25-332	22-341

4. Summary
In this work the co-pyrolysis of real waste HDPE and rice straw was investigated focussing the reaction kinetic parameters and long-term utilization of products. Based on TG result the apparent kinetic parameters were calculated using first-order kinetic approach. Significant differences were found among the decomposition of raw materials with 100%/0%, 70%/30%, 50%/50%, 30%/70% and 0%/100% plastic/biomass ratio. E.g., biomass decomposition had three stages, while HDPE only one. HDPE pyrolysis needs 272 kJ/mol activation energy, which can be significantly decreased by blending with biomass. Regarding corrosion and transportation stability, biomass derived pyrolysis oil showed non favourable properties, which was improved by plastic in raw material.

5. Acknowledgments
The authors acknowledge the National Office for Research and Innovation (NKTH), assistance granted under the Hungarian-Indian (KTIA-DST) R&D&I Program (TÉT_13_DST-1-2014-0003). We also acknowledge the financial support of Széchenyi 2020 under the EFOP-3.6.1-16-2016-00015.

6. References
[1] N. S. Shamsul, S. K. Kamarudin, N. A. Rahman. 2017 *Renew Sustain Ener* **80** 538-549
[2] S. M. Al-Salem, A. Antelava, A. Constantinou, G. Manos, A. Dutta 2017 *Journal of Environ. Manage* **197** 177-198
[3] R. Mianda, M. A. Barakat, S. Aburiazaiza Asad, M. Rehan, A. S. Nizami 2016 *Process Saf Emviron***102** 822-838
[4] N. Grioui, K. Halouani, F. A. Agblevor 2014 *Energy Sustain Dev* **21** 100-112
[5] J. V. Ortega, A. M. Renehan, M. W. Liberatore, A. M. Herring. 2011 *J Anal Appl Pyrol* **91** 190-19
[6] M. E. Boucher, A. Chaala, H. Pakdel, C. Roy 2000 *Biomass Bioenergy* **19** 351-361
[7] P. Brassard, S. Godbout, V. Raghavan. 2017 *Biosyst Eng* **161** 80-92
[8] P. Ranganathan, S. Gu 2016 *Bioresource Technol* **213** 333-341
[9] Y. Tang, Q. Huang, K. Sun, Y. Chi, J. Yan 2018 *Bioresource Technol* **249** 16-23
[10] A. Fernandez, A. Saffe, R. Pereyra, G. Mazza, R. Rodriguez. 2016, *Appl Therm Eng* **106** 1157-1164

IOP Conf. Series: Earth and Environmental Science **154** (2018) 012023 doi:10.1088/1755-1315/154/1/012023

Transient analysis of a solid oxide fuel cell stack with crossflow configuration

P Yuan and S F Liu

Mechanical Engineering Department, Lee-Ming Institute of Technology, 22 Sec. 3 Tailin Rd., Taishan 243, Taipei, Taiwan

Abstract. This study investigates the transient response of the cell temperature and current density of a solid oxide fuel cell having 6 stacks with crossflow configuration. A commercial software repeatedly solves the governing equations of each stack, and get the convergent results of the whole SOFC stack. The preliminary results indicate that the average current density of each stack is similar to others, so the power output between different stacks are uniform. Moreover, the average cell temperature among stacks is different, and the central stacks have higher temperature due to its harder heat dissipation. For the operating control, the cell temperature difference among stacks is worth to concern because the temperature difference will be over 10 °C in the analysis case. The increasing of the inlet flow rate of the fuel and air will short the transient state, increase the average current density, and drop the cell temperature difference among the stacks. Therefore, the inlet flow rate is an important factor for transient performance of a SOFC stack.

1. Introduction

The assembly of a solid oxide fuel cell (SOFC) includes an anode, a cathode, an electrolyte and an interconnector, as well as its operating temperature is between 600 and 1000°C. Because of the high operating temperature, the SOFC can apply the methane or ethanol as its fuel. The most shapes of a SOFC unit have two, cylindrical and flat shape, and the SOFC stack with flat shape is widely in application because this shape unit is easy for stacking. Due to the high operating temperature, a SOFC always combine other faculties to promote its power output, and the transient results of a SOFC stack become more important for the system control. This study surveys the transient research on the SOFC in the recent decade [1]-[10], and finds that all literature focusing on the transient analysis of a SOFC unit with parallel configuration because of its one-dimensional mathematical model. Most of these literature investigate the effect of parameters on the transient performance [1], [2]-[3], [7], [9], others develop the model for rapid calculating the transient results in the control viewpoint. The benefit of crossflow configuration is its easier inlet arrangement, and this configuration is generally applied in the SOFC stack. However, there is few literatures focusing on its transient analysis, especially for a SOFC stack. Therefore, the results of this study can provide a valuable information for the control of a SOFC operation.

2. Analysis

The schematic of a SOFC stack with crossflow configuration is shown in Fig. 1. This study considers that the gas in each channel is laminar and slug flow. Due to the repeating layers of SOFC stack, this study builds the governing equations for each layer, and then repeatedly calculates these stacking layers. In each layer, authors consider the anode, cathode, and electrolyte as a one component named the cell. Moreover, the variety in z direction is neglected because the length scale in thickness is smaller than that in the x and y direction, so this analysis is two-dimensional problem.

IOP Conf. Series: Earth and Environmental Science **154** (2018) 012023 doi:10.1088/1755-1315/154/1/012023

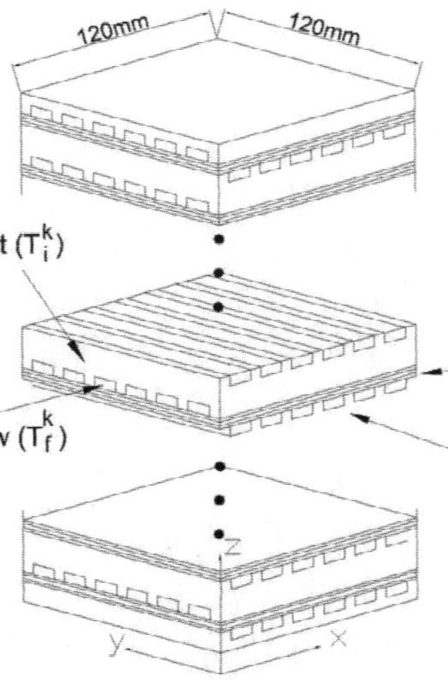

Figure 1. Schematic of a SOFC stack with crossflow configuration

The governing equations of each layer include the reaction equations, mass conservation equations, energy conservation equations, electrochemical equations.

2.1 Reaction equations

$$H_2 + O^{2-} \rightarrow H_2O + 2e^- \quad (1)$$

$$O_2 + 4e^- \rightarrow 2O^{2-} \quad (2)$$

2.2 Mass conservation equations

In the fuel side,

$$d(\frac{P}{RT_f} \cdot X_j^k)/dt + d\big(n_f^k \cdot X_j^k/l_y\big)/dx = S_j^k \quad (3)$$

In the air side,

$$d(\frac{P}{RT_a} \cdot X_j^k)/dt + d\big(n_a^k \cdot X_j^k/l_x\big)/dy = S_j^k \quad (4)$$

Meanwhile, $n_f = N_f/l_y$ and $n_a = N_a/l_x$. The fuel and air are ideal gas, and S_j stands for the reaction rate of reactants and products in the following:

$$S_{H_2} = -i/(2F), S_{H_2O} = i/(2F), S_{O_2} = -\,i/(4F) \quad (5)$$

2.3 Energy conservation equations

In the fuel side,

$$d(\frac{P}{RT_f} \cdot X_j^k \cdot c_{p,j}T_f^k)/dt + d\left(\sum_j \big(n_f^k/l_y \cdot X_j^k \cdot c_{p,j}T_f^k\big)\right)/dx$$

$$= (ha)_{if}\big(T_i^k - T_f^k\big) + (ha)_{cf}\big(T_c^k - T_f^k\big) + i/(4F) \cdot c_{p,O_3^{2-}}T_c^k \quad (6)$$

In the air side,

$$d(\frac{P}{RT_a} \cdot X_j^k \cdot c_{p,j}T_a^k)/dt + d\big(\Sigma_j\big(n_a^k/l_x \cdot X_j^k \cdot c_{p,j}T_a^k\big)\big)/dy = (ha)_{ia}\big(T_i^{k-1} - T_a^k\big) + (ha)_{ca}\big(T_c^k - T_a^k\big) - i/(4F) \cdot c_{p,O_3^{2-}}T_a^k \quad (7)$$

In the cell,

$$d(\rho_c\delta_c c_{p,c}T_c^k)/dt - (k\delta)_c\big(\partial^2 T_c^k/\partial x^2 + \partial^2 T_c^k/\partial y^2\big) =$$

$$(k)_{ic}(T_i^k - T_c^k)/\delta_{ic} + (k)_{ic}(T_i^{k-1} - T_c^k)/\delta_{ic} + (ha)_{cf}(T_f^k - T_c^k) + (ha)_{ca}(T_a^k - T_c^k) + i/(4F) \cdot c_{p,O_3^{2-}} T_a^k - i/(4F) \cdot c_{p,O_3^{2-}} T_c^k + \dot{q}_{reaction}^k \quad (8)$$

In the interconnector

$$d(\rho_i \delta_i c_{p,i} T_i^k)/dt - (k\delta)_i(\partial^2 T_i^k/\partial x^2 + \partial^2 T_i^k/\partial y^2) = (k)_{ic}(T_c^k - T_i^k)/\delta_{ic} + (k)_{ic}(T_c^{k+1} - T_i^k)/\delta_{ic} + (ha)_{if}(T_f^k - T_i^k) + (ha)_{ia}(T_a^{k+1} - T_i^k) \quad (9)$$

The subscript f is the fuel, a is the air, i is the interconnector, c is the cell, and the superscript k stands for the layer number. The initial, inlet, and boundary conditions are in the following:

$$\partial^2 T_c^k/\partial x^2 = \partial^2 T_c^k/\partial y^2 = 0,\ \partial^2 T_i^k/\partial x^2 = \partial^2 T_i^k/\partial y^2 = 0 \quad (10)$$

$$T_{initial} = T_{inlet}^{top} = T_{inlet}^{bottom} = T_{f,inlet} = T_{a,inlet} = T_0 \quad (11)$$

2.4 Nernst equation and polarization equations

$$E^k = V^k + V_{ohm}^k + V_{act}^k + V_{con}^k \quad (12)$$

$$E^k = E_0^k + (RT_c^k)/(2F) \ln\left(p_{H_2}^k p_{O_2}^{k^{0.5}} / p_{H_2O}^k\right) \quad (13)$$

$$E_0^k = 1.2723 - 2.7654 \times 10^{-4} T_c^k \quad (14)$$

$$V_{ohm}^k = i^k r_s^k \quad (15)$$

$$V_{act}^k = (RT_c^k)/(2F) \sinh^{-1}(i^k/(2i_{0,anode}^k)) + (RT_c^k)/(2F) \sinh^{-1}(i^k/(2i_{0,cathode}^k)) \quad (16)$$

$$V_{con}^k = -(RT_c^k)/(2F) \ln\left((1 - (RT_c^k)/(2F) \cdot (\delta_{anode}/D_{anode} p_{H_2}^k) \cdot i^k)/(1 + (RT_c^k)/(2F) \cdot (\delta_{anode}/D_{anode} p_{H_2O}^k) \cdot i^k)\right) \quad (17)$$

Meanwhile, $r_s = 300 \times 10^{-7}\ \Omega m^{-2}$、 $i_{0,anode} = 1290\ Am^{-2}$、 $i_{0,cathode} = 970\ Am^{-2}$、 $\delta_{anode} = 0.05mm$, $D_{anode} = 2 \times 10^{-5} ms^{-1}$ [11]。

3. Simulation

Authors have published some papers investigating the steady performance of the SOFC unit and stack by using both Fortran code developed by authors and the FlexPDE software [12]-[14], and have proved the reliability of this commercial software FlexPDE. Therefore, this study applies the software to solve the Eqns. (1) to (17) to get the unknown variables depending on the time and position, such as the molar fraction of the fuel and air, the temperature of fuel, air, cell, and interconnector, as well as the current density. For accuracy comparison, this study selects the same parameters of the analysis case in previous literature [13], [14] to run the steady and transient program for a SOFC unit. In the transient analysis program, the calculating time is from 0 to 1000s. Figure 2 and 3 show the current density and cell temperature distribution in the 200, 600s, and the steady state. In these figures, the color distribution represents the steady state by using the steady program, and the blue line distribution represents the transient result by using the transient program. The two figures indicate that the results of transient program match well with the results of steady program after 600s.

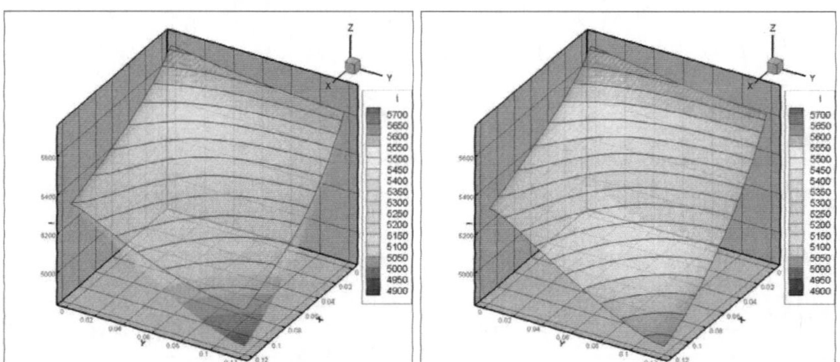

(a) 200s (b) 600s

Figure 2. Comparison of current density of a SOFC unit in transient and steady

IOP Conf. Series: Earth and Environmental Science **154** (2018) 012023 doi:10.1088/1755-1315/154/1/012023

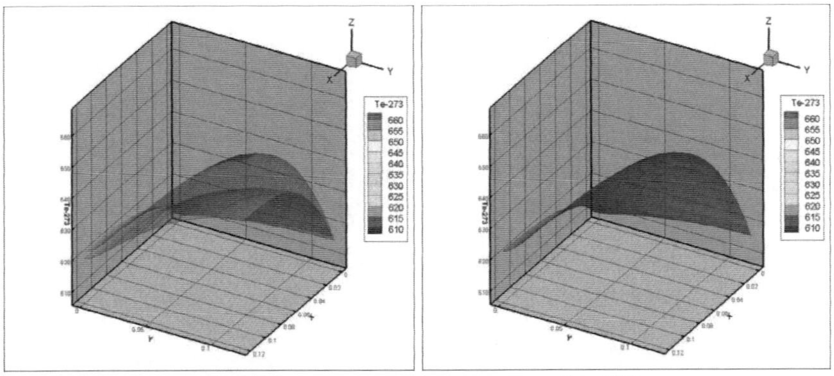

(a) 200s (b) 600s

Figure 3. Comparison of cell temperature of a SOFC unit in transient and steady

4. Results and Discussion

This study analyzes a SOFC stack with 6 layers, and the fuel composition is H_2 (50%) and CO_2(50%) as well as the air composition is O_2(21%) and N_2(79%). The inlet temperature of fuel and air is 900°C, and the molar flow rate of fuel and air inlet is 0.00023 and 0.01177 mol/s [15]. The operation voltage in this analysis case is 0.7V. Figure 4 shows the average current density of each layer from 0 to 600s. In this figure, it is obvious that the average current densities of every layer are similar, so the power output of each layer is uniform. Moreover, the SOFC stack gets the steady state at 600s in this analysis case.

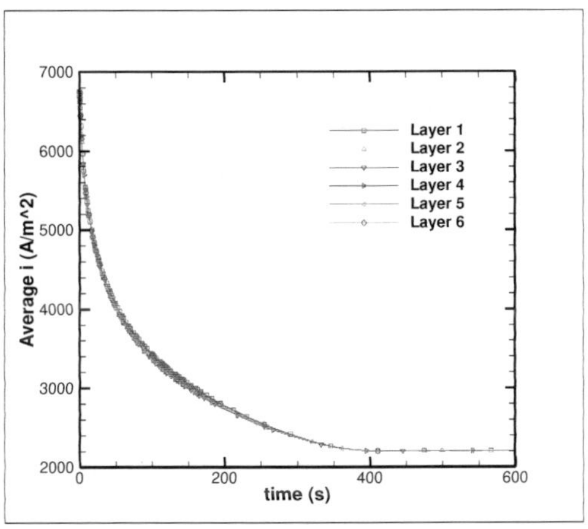

Figure 4. Response of average current density of each layer

Figure 5 shows the average cell temperature and cell temperature at the center of each layer from 0 to 600s. In Fig. 5(a), the layer 3 and 4 (blue lines) have the highest temperature, and the layer 1 and 6 (red lines) have the lowest temperature comparing to other layers. Because the reaction heat in the central layers is hard to spread out, the central layers must have higher temperature. In Fig. 5 (b), the temperature at cell center is slightly higher than that of average cell temperature of each layer, but the time getting steady state of temperature at cell center is shorter than that of average cell temperature. In Figs. 5(a) and 5(b), the temperature difference at different layers can be 12°C at near 100s, and be 8°C at steady state. Therefore, the cell temperature distribution in different layer of a SOFC stack in transient and steady state is worth to concerned.

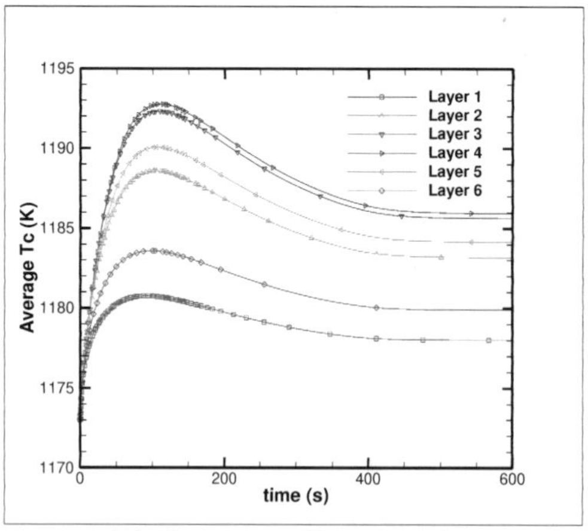

(a) Average cell temperature

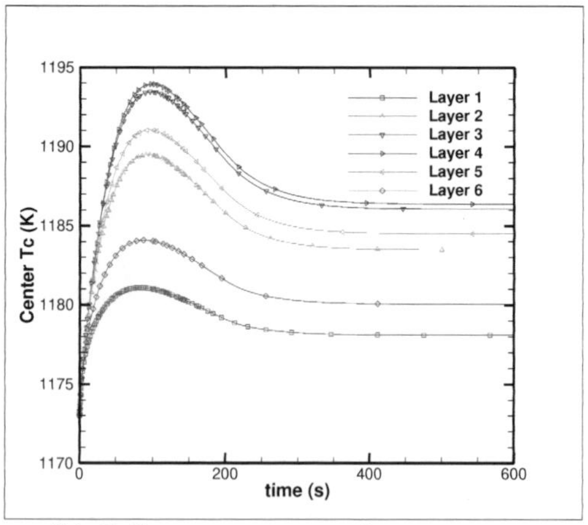

(b) Cell temperature at the center

Figure 5. Response of cell temperature of each layer

Figure 6 shows the average current density of the 3rd layer at one, two, and three times of original inlet molar flow rate of the fuel and air. In this figure, it is obvious that the increasing inlet molar flow rate will short the transient state, and increases the average current density. The time getting the steady state drops from 400s to less than 100s when the inlet flow rate is four times. Therefore, the change of inlet flow rate strongly affects the time getting the steady state, and it is an important factor in control viewpoint.

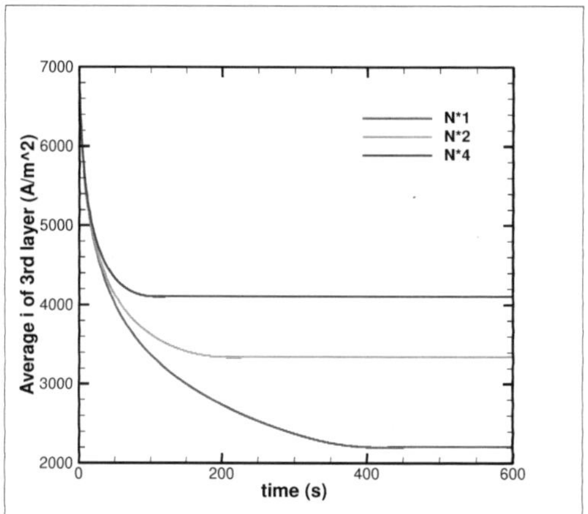

Figure 6. The average current densities of 3rd layer at different inlet molar flow rate.

Figure 7 depicts the average cell temperature of the 1st, 2nd, and 3rd layer in the SOFC stack at different inlet molar flow rates. Meanwhile, the continuous, dashed, and dash-dotted line represent the one, two, and four times of the original inlet flow rate of the fuel and air, respectively. The red, green, and blue line represents the 1st, 2nd, and 3rd layer of the SOFC stack. The results show that the increase of inlet molar flow rate not only shorts the transient state, but also drops the temperature difference among the layers. The increase of fuel and air can provide more reactants for the electrochemical reaction, as well as carry away more reaction heat from the cells. Although it seems that the increasing of fuel and air can promote the average current density in Fig. 6, and drops the cell temperature difference among the layers in Fig. 7; the fuel increasing is not a reasonable decision because of considering the fuel economy. However, the air increasing is a suitable method for decreasing the cell temperature difference in the SOFC stack.

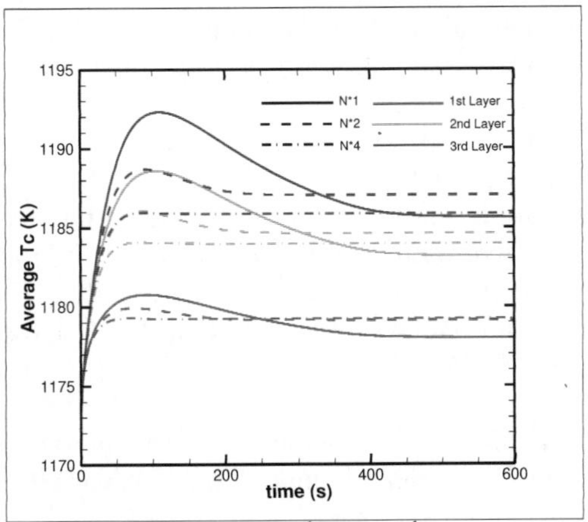

Figure 7. The average cell temperatures of 1st, 2nd, and 3rd layer at different inlet molar flow rate

5. Conclusions

This study applies a reliable commercial software, FlexPDE, to solve the mass equations, energy equations, and electrochemical equations of a SOFC with 6 stacks, as well as calculates the transient temperature and current density field of the whole stack. The preliminary results indicate that the average current density of each layer is similar to other layers, and the average cell temperature has obvious difference among different layers. Therefore, the cell temperature variation in stacks is worth to concerned in the control process. Moreover, the increasing of inlet flow rate will short the time to getting steady state, increase the average current density of each layer, and drop the cell temperature difference among the layers. Consequently, the inlet flow rate is an important factor affecting the transient performance of a SOFC stack.

6. References

[1] P. Iora, P. Aguiar, C. S. Adjiman, N. P. Brandon, Comparison of two IT DIR-SOFC models: Impact of variable thermodynamic, physical, and flow properties. Steady-state and dynamic analysis, *CHEM ENG SCI* 60 (2005) 2963–75.

[2] X. Xue, J. Tang, N. Sammes, Y. Du, Dynamic modeling of single tubular SOFC combining heat/mass transfer and electrochemical reaction effects, *J Power Sources* 142 (2005) 211–22.

[3] W. Jiang, R. X. Fang, J. A. Khan, R. A. Dougal, Parameter setting and analysis of a dynamic tubular SOFC model, *J Power Sources* 162 (2006) 316–26,

[4] H. Xi, J. Sun, V. Tsourapas, A control oriented low order dynamic model for planar SOFC using minimum Gibbs free energy method, *J Power Sources* 165 (2007) 253–66.

[5] Y. T. Qi, B. Huang, and J. L. Luo, 1-D Dynamic Modeling of SOFC with Analytical Solution for Reacting Gas-Flow Problem, *AIChE J* 54 (2008) 1537-53.

[6] Y. W. Kang, J. Li, G. Y. Cao, H. Y. Tu, J. Li, J. Yang, Dynamic temperature modeling of an SOFC using least squares support vector machines, *J Power Sources* 179 (2008) 683–92.

[7] X. Zhang, Y. M. Wu, A Control-oriented Dynamic Model Adapted to Variant Steam-to-carbon Ratios for an SOFC with Exhaust Fuel Recirculation, *FUEL CELLS* 11 (2011) 200–11.

[8] X. J. Wu, Q. Huang, X. J. Zhu, Dynamic modeling of a SOFC/MGT hybrid power system based on modified OIF Elman neural network, *Int J Energy Res* 36 (2012) 87–95.

[9] S. A. Hajimolan, M. A. Hussain, A.W. Daud, M. H. Chakrabarti, Dynamic modelling and sensitivity analysis of a tubular SOFC fuelled with NH_3 as a possible replacement for H_2, *CHEM ENG RES DES* 90 (2012) 1871-82.

[10] A. S. Martinez, J. Brouwer, G. S. Samuelsen, Feasibility study for SOFC-GT hybrid locomotive power: Part I. Development of a dynamic 3.5MW SOFC-GT FORTRAN model, *J Power Sources* 213 (2012) 203-17.

[11] S. H. Chan, K. A. Khor, Z. T. Xia, A complete polarization model of a solid oxide fuel cell and its sensitivity to the change of cell component thickness, *J Power Sources* 93 (2001) 130-40.

[12] P. Yuan and S. F. Liu, Numerical Analysis of Temperature and Current Density Distribution of a Planar Solid Oxide Fuel Cell Unit with Non-uniform Inlet Flow, *NUMER HEAT TR A-APPL* 51 (10) (2007) 941-57.
[13] P. Yuan, Effect of inlet flow maldistribution in the stacking direction on the performance of a solid oxide fuel cell stack, *J Power Sources* 185 (2008) 381-91.
[14] P. Yuan, S. F. Liu, Effect of non-uniform inlet flow rate on the heat-up process of a solid oxide fuel cell unit with cross-flow configuration, *Int J Hydrogen Energy* 41 (2016) 12377-86.
[15] O. Razbani, M. Assadi, M. Andersson, Three dimensional CFD modeling and experimental validation of an electrolyte supported solid oxide fuel cell fed with methane-free biogas, *Int J Hydrogen Energy* 38 (2013) 10068-80.

Acknowledgments

The authors would like to thank the Ministry of Science and Technology of the Republic of China, Taiwan, for the financial support of this research under Contract No. MOST- 106- 2221- E- 234- 002.

Institute of Physics
Dirac House, Temple Back
Bristol BS1 6BE UK

ISSN: 1755-1307
ISBN 978-1-5108-6337-8

9 781510 863378

CURRAN ASSOCIATES INC.
proceedings.com

II International Workshop "Thermal Methods for Enhanced Oil Recovery: Laboratory Testing, Simulation and Oilfields Applications" (ThEOR 2017)

IOP Conference Series: Earth and Environmental Science
Volume 155

Kazan, Russia
19 – 23 June 2017

IOP
Institute of Physics